Mitochondrial Medicine

A Primer for Health Care Providers and Translational Researchers

Mitochondrial Medicine

A Primer for Health Care Providers and Translational Researchers

PANKAJ PRASUN, MD
Assistant Professor
Division of Medical Genetics
Department of Genetics and Genomic Sciences
Icahn School of Medicine at Mount Sinai
One Gustave L. Levy Place
New York, NY, United States

ELSEVIER

ACADEMIC PRESS
An imprint of Elsevier

Publisher: Andre Gerhard Wolff
Acquisition Editor: Peter B. Linsley
Editorial Project Manager: Karen Miller
Production Project Manager: Sreejith Viswanathan
Cover Designer: Matthew Limbert

Academic Press is an imprint of Elsevier
125 London Wall, London EC2Y 5AS, United Kingdom
525 B Street, Suite 1650, San Diego, CA 92101, United States
50 Hampshire Street, 5th Floor, Cambridge, MA 02139, United States
The Boulevard, Langford Lane, Kidlington, Oxford OX5 1GB, United Kingdom

Working together
to grow libraries in
developing countries

www.elsevier.com • www.bookaid.org

This book is dedicated to all the patients with mitochondrial disease and their families.

Preface

Mitochondrial medicine is one of the most rapidly advancing branches of medicine. This decade has shown an exponential rise in the number of diseases attributed to mitochondria due to availability and widespread use of whole exome- and genome-sequencing techniques. Mitochondrial diseases are relatively common. Prevalence of mitochondrial disease due to mutations of mitochondrial DNA alone is estimated to be 1 in 5000. Considering most of the mitochondrial diseases are caused by nuclear gene mutations, the overall prevalence is even higher. Unfortunately, a curative treatment is not available for most of the mitochondrial diseases at present. Hence, a better understanding of mitochondrial physiology and intensive research is needed to address the need of an ever-increasing patient population. Traditionally, mitochondria have been labeled as the "powerhouse of the cell" due to its vital role in energy production. However, it is now recognized that mitochondria also play vital roles in many other physiological processes such as iron and calcium homeostasis, apoptosis, cell signaling, and biosynthesis of macromolecules. Disorders of mitochondria are no longer limited to "energy failure" conditions, rather mitochondria are increasingly being recognized as key players in more common disorders such as neurodegenerative diseases and cancer. It is even more critical now to understand mitochondrial physiology to make strides in the prevention and treatment of these disorders. This book is an attempt to summarize the diverse roles of mitochondria in human health and disease. It has been divided into three sections. Section 1 gives an overview of clinical presentation, diagnosis, and management of inherited mitochondrial diseases. Section 2 outlines the types of inherited mitochondrial diseases. Clinical presentation and management of the most commonly seen mitochondrial diseases are described in detail. Section 3 describes the role of mitochondria in common diseases such as nonalcoholic fatty liver disease, Parkinson disease, and cancer. Pathogenesis of these conditions from the mitochondrial perspective is outlined. The purpose is to examine the role of mitochondria in these conditions and not to replace the traditional understanding of those conditions. My hope is that clinicians, as well as researchers, will find this book as a useful quick read to get an overview of this very fascinating, rapidly evolving branch of medicine.

Sincerely,
Pankaj Prasun, MD
New York, USA

About This Book

Mitochondrial medicine is a rapidly evolving and fascinating branch of medicine. The role of mitochondria in human health and disease is increasingly being recognized. Mitochondrial dysfunction is not only implicated in rare inherited disorders but is also a significant contributor to the pathogenesis of common medical conditions such as stroke and cancer. It is an area of intense clinical and basic science research. *Mitochondrial Medicine: A Primer for Healthcare Providers and Translational Researchers* is intended to provide an overview of this multidisciplinary branch of medicine. This book is divided into three sections comprising 21 chapters through which **Dr. Pankaj Prasun** provides a broad overview of clinical presentation, laboratory diagnosis, and treatment of mitochondrial diseases. The most common mitochondrial diseases are described separately for a quick bedside reference. In addition, the role of mitochondrial dysfunction in common medical conditions such as obesity, type 2 diabetes, and heart failure has been explored. While providing a solid foundation in its topic area, each chapter in *Mitochondrial Medicine* is concise and written in an accessible format for a quick reference at the patient's bedside or in the clinical laboratory.

KEY FEATURES

- Provides broad introduction of mitochondria and their dysfunction in human health and disease.
- Current practice of diagnosis, treatment, genetic counseling, and prenatal testing of mitochondrial diseases are described.
- Broad spectrum of inherited mitochondrial disorders is organized in chapters according to their pathogenesis for easy understanding.
- More commonly encountered mitochondrial diseases are described separately as subchapters.
- Provides thorough coverage of inherited mitochondrial disorders, as well as the role of mitochondria, in common medical conditions.
- Features short, accessible chapters for quick reference at the patient's bedside or in the clinical laboratory.

Abbreviation List

3-MGA	3-Methyl Glutaconic Aciduria	MELAS	Mitochondrial Encephalomyopathy Lactic Acidosis, and Stroke-Like episodes
AAV	Adeno-Associated Virus		
AD	Alzheimer's Disease	MERRF	Myoclonic Epilepsy Ragged-Red Fibers
ADP	Adenosine Diphosphate		
AICAR	Aminoimidazole Carboxamide Ribonucleoside	MFF	Mitochondrial Fission Factor
		MIDD	Maternally Inherited Diabetes and Deafness
ALS	Amyotrophic Lateral Sclerosis		
AMP	Adenosine Monophosphate	MILS	Maternally Inherited Leigh Syndrome
AMPK	Adenosine Monophosphate-Activated Protein Kinase	MMP	Mitochondrial Membrane Potential
		MNGIE	Mitochondrial Neurogastrointestinal Encephalopathy
ANT	Adenine Nucleotide Translocator		
AOX	Alternative Oxidase	MOMP	Mitochondrial Outer Membrane Permeabilization
ATP	Adenosine Triphosphate		
CMT	Charcot-Marie-Tooth	MRI	Magnetic Resonance Imaging
CoQ	Coenzyme Q	MRS	Magnetic Resonance Spectroscopy
CPEO	Chronic Progressive External Ophthalmoplegia	mtDNA	Mitochondrial Deoxyribonucleic Acid
		MTS	Mitochondrial-Targeting Sequence
CVD	Cardiovascular Diseases	NAC	N-Acetyl Cysteine
CVS	Chorionic Villus Sampling	NAD	Nicotinamide Adenine Dinucleotide
DCA	Dichloroacetate	NAFLD	Nonalcoholic Fatty Liver Disease
DCM	Dilated Cardiomyopathy	NARP	Neurogenic muscle weakness Ataxia Retinitis Pigmentosa
DCMA	Dilated Cardiomyopathy with Ataxia		
DGUOK	Deoxyguanosine Kinase	NASH	Nonalcoholic Steatohepatitis
DLD	Dihydrolipoamide Dehydrogenase	NLRP3	Nod-Like Receptor Protein 3
DM	Diabetes Mellitus	NO	Nitric Oxide
DNA	Deoxyribonucleic Acid	NPO	Nil Per Os (nothing by mouth)
DRP1	Dynamin-Related Protein 1	NRF	Nuclear Respiratory Factor
EE	Ethylmalonic Encephalopathy	OMM	Outer Mitochondrial Membrane
EEG	Electro Encephalogram	OPA	Optic Atrophy
ETC	Electron Transport Chain	OXPHOS	Oxidative Phosphorylation
FA	Friedreich Ataxia	PA	Phosphatidic Acid
FAD	Flavin Adenine Dinucleotide	PC	Pyruvate Carboxylase
HCM	Hypertrophic Cardiomyopathy	PCR	Polymerase Chain Reaction
HD	Huntington Disease	PD	Parkinson's Disease
HDL	High Density Lipoprotein	PDH	Pyruvate Dehydrogenase
HSP	Heat Shock Protein	PEO	Progressive External Ophthalmoplegia
IMM	Inner Mitochondrial Membrane	PGC1α	Peroxisome proliferator activated receptor Gamma Coactivator 1α
ISC	Iron Sulfur Cluster		
KSS	Kearns-Sayre Syndrome	PIGD	Preimplantation Genetic Diagnosis
LDL	Low-Density Lipoprotein	POLG	Polymerase Gamma
LHON	Leber Hereditary Optic Neuropathy	PPAR	Peroxisome Proliferator Activated Receptor
LS	Leigh Syndrome		
MCU	Mitochondrial Calcium Uniporter	PTP	Permeability Transition Pore

RCM	Restrictive Cardiomyopathy	TCA	Tricarboxylic Acid
RNA	Ribonucleic Acid	TIM	Translocase of Inner Membrane
RNS	Reactive Nitrogen Species	TK	Thymidine Kinase
ROS	Reactive Oxygen Species	TOM	Translocase of Outer Membrane
SANDO	Sensory Ataxia Neuropathy Dysarthria Ophthalmoparesis	TPP	Thiamine Pyrophosphate
		VLDL	Very Low-Density Lipoprotein
SDH	Succinate Dehydrogenase		
TALEN	Transcription Activator-Like Effectors Nucleases		

Notice

Medicine is an ever-changing science. As clinical experience and new research increase our knowledge, changes/modifications in treatment are required. The author and the publisher of this work have checked with sources believed to be reliable in their efforts to provide information that is complete and in accord with the standards accepted at the time of publication. However, in view of the possibility of human error or changes in medical sciences, neither the authors nor the publisher nor any other party who has been involved in the preparation or publication of this book warrants that the information contained herein is in every respect accurate or complete, and they disclaim all responsibility for any errors or omissions or for the results obtained from use of the information contained in this book. Readers are encouraged to confirm the information contained herein with other sources that changes have not been made in the recommended dose or in the contraindications for administration.

Contents

SECTION I
BASICS OF MITOCHONDRIAL MEDICINE

1 Functions of Mitochondria, *1*

2 Clinical Presentation of Mitochondrial Diseases, *5*

3 Mitochondrial Genetics, *7*

4 Diagnosis of Mitochondrial Diseases, *11*

5 Treatment of Mitochondrial Diseases, *15*

6 Management of Emergency and Surgery, *21*

7 Genetic Counseling, Prenatal Diagnosis, and Reproductive Options in Mitochondrial Diseases, *23*

SECTION II
INHERITED MITOCHONDRIAL DISEASES

8 Mitochondrial Diseases: Classification, *27*

9 Mitochondrial DNA Deletion Disorders, *31*

10 Mitochondrial DNA Point Mutation Disorders, *37*

11 Mitochondrial Disease of Nuclear Origin: Respiratory Chain Complex Units, *49*

12 Mitochondrial Disease of Nuclear Origin: Disorders of Mitochondrial DNA Replication, *59*

13 Mitochondrial Disease of Nuclear Origin: Disorders of Mitochondrial DNA Transcription and Translation, *67*

14 Mitochondrial Disease of Nuclear Origin: Disorders of Mitochondrial Homeostasis, *77*

15 Disorders of Pyruvate Metabolism and Tricarboxylic Acid Cycle, *83*

SECTION III
MITOCHONDRIA AND COMMON MEDICAL CONDITIONS

16 Mitochondria and Aging, *97*

17 Mitochondria, Obesity, Metabolic Syndrome, and Type 2 Diabetes, *101*

18 Mitochondria and Nonalcoholic Fatty Liver Disease, *103*

19 Mitochondria and Cardiovascular Diseases, *105*

20 **Mitochondria and Late-Onset Neurodegenerative Diseases,** *109*

21 **Mitochondria and Cancer,** *113*

USEFUL WEBSITES, *115*

COMMONLY PRESCRIBED MITOCHONDRIAL SUPPLEMENTS, *117*

INDEX, *119*

CHAPTER 1

Functions of Mitochondria

ABSTRACT

Mitochondrial medicine is one of the most rapidly advancing branches of medicine. Apart from rare inherited mitochondrial diseases caused by genetic changes, mitochondrial dysfunction is implicated in common health problems such as obesity and cancer. In this chapter, the basic functions of mitochondria are described to understand its role in human health and disease.

KEYWORDS

Apoptosis; ATP; Electron transport chain; Oxidative phosphorylation; Reactive oxygen species; Respiratory chain complex.

POWERHOUSE OF THE CELL

Mitochondria are small organelles inside the cell. They have double membrane structure—outer mitochondrial membrane (OMM) and inner mitochondrial membrane (IMM). These are separated by intermembranous space. Mitochondria are also called powerhouse of the cell because they are the main sites of ATP (energy currency of the cell) production. The main source of energy, glucose, and fat are oxidized inside mitochondria via tricarboxylic acid (TCA) cycle and beta-oxidation, respectively. The energy stored as chemical bonds in food is released by these processes as high energy electrons, which are captured by NAD+ and FAD (which are then reduced to NADH and FADH$_2$). NADH and FADH$_2$ molecules donate these high energy electrons to electron transport chain (ETC), which is located in the IMM. As the electron travels down this chain from high energy to low energy state, the dissipated energy is harnessed to propel proton from inside of mitochondria called mitochondrial matrix across the IMM to the intermembranous space. A proton gradient is thus generated. Finally, the proton moves down its concentration gradient across the inner membrane from intermembranous space to mitochondrial matrix. This movement dissipates energy that is harnessed to form ATP molecules. More than 90% of total cellular ATP is produced in mitochondria. Thus, mitochondria play a vital role in the energy metabolism.

ELECTRON TRANSPORT CHAIN

ETC is located in the IMM. It consists of five complexes: complexes I–V. Complex I receives electrons from the NADH; hence, it is also called NADH dehydrogenase. Complex II receives electrons from FADH$_2$ molecule. Complex II is also a part of TCA cycle where it catalyzes the conversion of succinate to fumarate by its succinate dehydrogenase activity. Complexes I and II donate electrons to coenzyme Q (CoQ). CoQ can freely diffuse through the IMM to complex III. Complex III accepts electrons from the CoQ and reduces cytochrome c (part of complex III) by its cytochrome c reductase activity. Cytochrome c is oxidized by complex IV (cytochrome c oxidase) by accepting electrons from it and passing it to oxygen that results in the formation of water. As the electron moves along this chain, free energy is released. It is used to pump protons at complexes I, III, and IV from the mitochondrial matrix to the intermembranous space generating a proton gradient. Protons then diffuse along its concentration gradient at complex V or ATP synthase where energy generated by the proton drive across IMM is used to generate ATP from ADP (Fig. 1.1).

OTHER ROLES OF THE MITOCHONDRIA

Apart from ATP production, fatty acid oxidation, and TCA cycle, several other metabolic processes occur inside mitochondria such as ketogenesis, parts of urea cycle, heme synthesis, and phospholipid synthesis. New roles of mitochondria in health and disease are emerging. Some of these are briefly discussed here.

Mitochondria and Reactive Oxygen Species

ETC plays a vital role in the energy metabolism of cells by passing high energy electrons to oxygen in a step-wise

Mitochondrial Medicine. https://doi.org/10.1016/B978-0-12-817006-9.00001-0

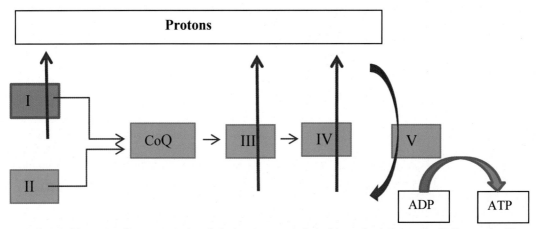

FIG. 1.1 Diagrammatic representation of electron transport chain. I: complex I, II: complex II, III: complex III, IV: complex IV, V: complex V. Electron movement is shown by *red arrows*; proton movement is shown by *purple arrows*.

fashion. However, high energy electrons may slip directly to oxygen rather than following this controlled path. When this happens, superoxide radicals are generated. These superoxide radicals then form hydrogen peroxide rather than water. Hydrogen peroxide can produce hydroxyl radicals that are highly reactive and dangerous. Hydroxyl radicals can damage membranes, proteins, enzymes, and DNA, which may result in cell death. However, mitochondria also have a very efficient antioxidant mechanism. It consists of superoxide dismutase that converts superoxide radicals to hydrogen peroxide, which is then converted to water by the enzyme glutathione peroxidase in the presence of reduced glutathione. Thus, the production of hydroxyl radical is minimized. This system is very effective and scavenges most of the reactive oxygen species (ROS) produced locally in the mitochondria. Moreover, mitochondria also play a significant role in scavenging ROS from other cellular sources such as peroxisome. Mitochondrial dysfunction may lead to excessive ROS generation and cell injury. Increased ROS due to mitochondrial dysfunction as a pathophysiological mechanism of neurodegenerative diseases and aging is increasingly being recognized.

Mitochondria and Calcium Homeostasis

The cellular processes associated with increased energy demand such as contraction, secretion, and movement are usually associated with increased cytoplasmic calcium. An increase in cytoplasmic calcium leads to increased mitochondrial calcium where it activates the regulatory enzymes of the TCA cycle and enhances the function of the ATP synthase component of the ETC.

Thus, increased demand of energy is met by increased energy production from the mitochondria. Calcium entry inside mitochondria is promoted by the negative electric potential of the mitochondria matrix and facilitated by calcium-selective channel on the IMM called mitochondrial calcium uniporter. Calcium exchangers extrude calcium from the mitochondria in exchange of sodium or proton entry thus bringing the calcium level inside mitochondria to basal level. Calcium is a very important intracellular signaling molecule modulating many important biophysical processes. Hence, its level is tightly controlled inside the cell. Calcium is actively pumped out of the cell and inside the endoplasmic reticulum to maintain a low intracellular concentration. In conditions such as hypoxia, these energy-dependent processes fail leading to high intracellular calcium that leads to high intramitochondrial calcium. Sustained high intramitochondrial calcium is detrimental as it leads to the formation of permeability transition pore (PTP) in the mitochondrial membrane with subsequent mitochondrial swelling, energy failure, and ultimately cell death.

Mitochondria in Iron and Copper Metabolism

Heme, iron—sulfur (Fe—S) clusters, and copper (Cu) are essential components of mitochondrial ETC. Heme, along with Fe—S clusters and Cu-bound cofactors, participates in electron transfer. Heme is present as a prosthetic group in complex II, which is essential for its function. It is present in cytochrome *bc*1 complex of complex III where it mediates electron transfer. The catalytic site of complex IV (cytochrome *c* oxidase) contains two heme moieties and two Cu centers that accept electrons and transfer it to oxygen. Fe—S clusters mediate electron transfer in complexes I, II, and III. Because of the vital role of heme, Fe—S clusters, and Cu in ETC function, disorders of their synthesis or incorporation into the ETC can lead to mitochondrial disease. For example, deficiency of ISCU, a scaffold protein for Fe—S cluster biogenesis, leads to myopathy and lactic acidosis. Heme synthesis occurs by iron incorporation into protoporphyrin IX by ferrochelatase inside mitochondria. Mitochondria are also the site for Fe—S cluster biogenesis. When synthesis of Fe cofactors is decreased in mitochondria, compensatory mechanisms lead to increased cellular iron uptake and subsequently mitochondrial iron overload that can cause mitochondrial dysfunction by excessive ROS generation. A decrease in Fe—S cluster biosynthesis also leads to a secondary deficiency of heme synthesis. Heme is an essential component of hemoglobin. Sideroblastic anemia characterized by anemia and mitochondrial iron overload in erythroid precursors can be due to a primary disorder of heme synthesis, or secondary to a Fe—S biogenesis disorder, or due to inherited and acquired mitochondrial dysfunction affecting mitochondrial iron metabolism.

Mitochondria in Immunity and Inflammation

The outer mitochondrial membrane provides a platform for innate immunity signaling. The single- and double-stranded RNAs from viral infection are sensed by retinoic acid-inducible protein-like receptor family, which then activates mitochondrial antiviral signaling protein on OMM resulting in the production of interferons and proinflammatory cytokines. In addition, various pathogen-associated molecular pattern and damage-associated molecular pattern (DAMP) are recognized by Toll-like receptors and Nod-like receptors (NLR) leading to the production of proinflammatory cytokines. Upon activation, NLRs form multiprotein complexes called inflammasomes. NLRP3 (Nod-like receptor protein 3) is an important member of the NLR family, as it is activated by stimuli of diverse origin. Mitochondria resemble bacteria in certain ways. Mitochondrial DNA and N-formyl peptide, when released in the cytoplasm, can act as DAMPs. Mitochondrial dysfunction may lead to excessive production of ROS and release of mtDNA in the cytoplasm. Both can lead to NLRP3 activation and induction of chronic inflammation. Mitochondrial dysfunction contributes to chronic inflammation seen in major health problems such as obesity, metabolic syndrome, and nonalcoholic steatohepatitis.

Mitochondria and Cell Death

Cell death is of two types: programmed cell death or apoptosis, and accidental cell death or necrosis. Mitochondria play a significant role in both types of cell death. Apoptosis is an important physiological process, which is required for fetal development and also the removal of unwanted and damaged cells later in life. The most important mechanism of apoptosis is the activation of a group of proteases called caspases. Caspases target numerous substrates leading to apoptosis. Apoptosis can be initiated by two pathways: intrinsic and extrinsic. Apoptosis by intrinsic pathway can be initiated by different stimuli such as DNA damage, ROS injury, hypoxia, lack of hormones, or growth factors. The ultimate outcome of these stimuli is mitochondrial outer membrane permeabilization (MOMP) and release of proapoptotic factors such as cytochrome *c* and apoptosis-inducing factor from the intermembranous space to the cytoplasm. Cytochrome *c* activates a cascade of reactions in the cytoplasm leading to caspase 3 activation resulting in apoptosis. The extrinsic pathway is initiated by binding of death signals to extracellular receptor and activation of the intracellular pathway leading to caspase 8 activation. Caspase 8 initiates apoptosis by several mechanisms including activation of caspase 3, and MOMP. Apoptosis is a coordinated and energy-dependent process. However, necrosis is because of energy failure. Conditions such as hypoxia and reperfusion injury in heart muscle cells lead to excessive intramitochondrial calcium and ROS production. These lead to the creation of PTP in IMM that causes the collapse of mitochondrial membrane potential, ATP depletion, mitochondrial swelling, and rapid cell death (necrosis).

SUGGESTED READING

1. Nunnari J, Suomalainen A. Mitochondria: in sickness and in health. *Cell.* 2012;148:1145—1159.
2. Venditti P, Di Stefano L, Di Meo S. Mitochondrial metabolism of reactive oxygen species. *Mitochondrion.* 2013;13: 71—82.
3. Giorgi C, Agnoletto C, Bononi A, et al. Mitochondrial calcium homeostasis as potential target for mitochondrial medicine. *Mitochondrion.* 2012;12:77—85.
4. Xu W, Barrientos T, Andrews NC. Iron and copper in mitochondrial diseases. *Cell Metab.* 2013;17:319—328.
5. Mills EL, Kelly B, O'Neill LAJ. Mitochondria are the powerhouses of immunity. *Nat Immunol.* 2017;18:488—498.
6. Dawson TM, Dawson VL. Mitochondrial mechanisms of neuronal cell death: potential therapeutics. *Annu Rev Pharmacol Toxicol.* 2017;57:437—454.

CHAPTER 2

Clinical Presentation of Mitochondrial Diseases

ABSTRACT

Diagnosis of a mitochondrial disorder can be very challenging due to a wide spectrum of clinical manifestations. It can affect any organ system and may present at any age. Hence, awareness of different clinical manifestations of mitochondrial disease is necessary to suspect a mitochondrial condition. This chapter provides a broad overview of clinical presentations of mitochondrial diseases.

KEYWORDS

Encephalomyopathy; Mitochondrial disease; Multisystem disease; Neuromuscular disorder.

MITOCHONDRIAL DISEASE VERSUS MITOCHONDRIAL DYSFUNCTION

As noted in the previous chapter, mitochondria play diverse metabolic and nonmetabolic roles. However, "mitochondrial disease" is defined as a group of conditions caused by impairment of mitochondrial respiratory (electron transport) chain. These are genetic disorders that affect one or more components of the respiratory chain also called oxidative phosphorylation or OXPHOS system. Mitochondrial dysfunction is a more generalized term. Mitochondrial function may be impaired by inherited, acquired, or environmental conditions via several mechanisms. For example, mitochondrial dysfunction is implicated in the pathophysiology of many neurodegenerative disorders, aging, diabetes, metabolic syndrome, and cancer. However, these disorders are not typically labeled as "mitochondrial diseases" as mitochondrial dysfunction reflects only one of the several aspects of their pathophysiology.

CLINICAL FEATURES OF MITOCHONDRIAL DISEASES

Mitochondria have a central role in the energy production for the cell. Hence, any cell that depends on oxidative phosphorylation for energy can be affected by mitochondrial disease. The organ systems with high energy requirement such as brain, heart, and muscle are more often and more severely affected. The mitochondrial disease often affects more than one organ systems. However, mitochondrial disease can present with only one organ system symptoms. Following are the main clinical features of mitochondrial diseases:

Central Nervous system: Seizures, developmental delay, neuroregression, migraine, ataxia, stroke.

Peripheral nervous system: Neuropathy.

Eye: Optic neuropathy, retinopathy, ophthalmoplegia.

Ear: Sensorineural hearing loss.

Heart: Cardiomyopathy, arrhythmia.

Muscle: Fatigue, exercise intolerance, myopathy

Gastrointestinal system: Pseudoobstruction, delayed gastric emptying.

Liver: Hepatopathy.

Endocrine system: Diabetes, hypothyroidism, hypoparathyroidism, hypogonadism.

Kidney: Renal Fanconi syndrome, glomerulopathy.

Blood: Anemia, pancytopenia, neutropenia.

WHEN TO SUSPECT A MITOCHONDRIAL DISEASE

The clinical spectrum of mitochondrial disease is wide and expanding. Many mitochondrial diseases such as Mitochondrial Encephalopathy, Lactic Acidosis, and Stroke-like episodes (MELAS) can be identified based on characteristic constellation of features. However, often all features are not present. Hence, a mitochondrial disease should be suspected whenever there is multisystem involvement without a unifying diagnosis.

SUGGESTED READING

1. Niyazov DM, Kahler SG, Frye RE. Primary mitochondrial disease and secondary mitochondrial dysfunction: importance of distinction for diagnosis and treatment. *Mol Syndromol.* 2016;7:122−137.
2. Zeviani M, Di Donato S. Mitochondrial disorders. *Brain.* 2004;127:2153−2172.

Mitochondrial Medicine. https://doi.org/10.1016/B978-0-12-817006-9.00002-2

CHAPTER 3

Mitochondrial Genetics

ABSTRACT

Mitochondria are unique in that they have their own genetic material. However, they are not autonomous. Mitochondrial genetic material is incomplete. Nuclear genes are required for mitochondrial biogenesis and mitochondrial functions. The understanding of mitochondrial genetics is not only important to understand the inheritance of mitochondrial diseases but also vital to understand pathophysiology, diagnosis, and treatment of mitochondrial disorders.

KEYWORDS

Genetics; Heteroplasmy; Maternal inheritance; Mitochondrial DNA; Threshold effect.

MITOCHONDRIAL GENOME

The mitochondrial genome is a circular double-stranded DNA molecule of about 16.6 kb size. It has 37 genes: 13 genes encode a component of electron transport chain, 22 encode mitochondrial t-RNAs, and 2 encode mitochondrial ribosomal RNAs. Electron transport chain complexes are composed of many subunits. Mitochondrial DNA (mtDNA) encodes only part of the electron transport chain (ETC). The remaining subunits are encoded by nuclear genes (Table 3.1).

Replication of Mitochondrial DNA

Mitochondrial DNA is constantly replicating. The replication starts with the unwinding of the double strand by helicase (*TWNK*). After this, a single strand is stabilized by mitochondrial single-strand binding protein (*SSBP1*). Polymerase complex consisting of one polymerase gamma (*POLG1*) and two accessory subunits encoded by *POLG2* replicate mitochondrial DNA.

Transcription of Mitochondrial DNA

Each strand of the mitochondrial DNA is transcribed as one pre-RNA molecule. This is carried out by the mitochondrial RNA polymerase (*POLRMT*) in the presence of one of the two transcription factors, TFAM or TFB2M. A mature transcript is terminated by mitochondrial termination factor 1 (*MTERF1*).

Translation of Mitochondrial DNA

Mitochondrial DNA encodes for the complete set of t-RNAs. However, they need to be processed before participating in the protein synthesis. Aminoacyl t-RNA synthases are a group of enzymes that attach specific amino acids to their corresponding t-RNAs. For example, arginyl t-RNA synthase (*RARS2*) conjugates arginine with its corresponding t-RNA. Methionine t-RNA formyltransferase (*MTFMT*) catalyzes formylation of the methionine attached to the t-RNA^{met} required for translation initiation. Pseudouridine synthase (*PUS1*) causes pseudouridylation necessary to facilitate base pairing for t-RNA secondary structure. Translation occurs at the mitochondrial ribosome. Mitochondrial ribosome consists of two subunits; the smaller subunit consists of 12S r-RNA and 30 ribosomal proteins while the larger subunit is made of 16S r-RNA and 40 ribosomal proteins.

Mitochondrial translation starts with initiation complex formation that consists of the smaller ribosomal subunit, m-RNA, formyl methionine t-RNA^{met}, and mitochondrial translation initiation factor 2 and 3 (*MTIF2*/*MTIF3*). When a start codon is encountered, formyl methionine t-RNA binds the codon, and the translation is started. The elongation process is carried out by aminoacyl t-RNAs in the presence of mitochondrial elongation factors until a stop codon is encountered when mitochondrial translation release factor (*MTRF1L*) causes termination of translation.

Other than m-RNA, t-RNAs and r-RNAs, all other components of mtDNA replication, transcription, and translation are encoded by nuclear genes (*shown in parentheses*).

HETEROPLASMY, THRESHOLD EFFECT, AND SOMATIC SEGREGATION

There are thousands of mitochondria in a single cell. Hence, a single cell may harbor mtDNA with pathogenic mutation and wild-type mtDNA together. This is called heteroplasmy and measured as the percentage of mtDNA with mutation. Heteroplasmy level may vary between different cells of a tissue, between different tissues of an individual, and between

TABLE 3.1
Genetic Origin of Electron Transport Chain Complexes

Complex	Subunits	Nuclear Encoded	mtDNA Encoded
I	44	37	7
II	4	4	0
III	11	10	1
IV	14	11	3
V	19	17	2

different individuals carrying the same mtDNA mutation. A higher heteroplasmy may lead to clinical symptoms while a low level of heteroplasmy may be clinically silent. Typically, clinical and/or biochemical manifestations from a pathogenic mtDNA mutation are seen when heteroplasmy level exceeds 60%–80%. This is called "threshold effect" However, the threshold varies between different tissues and for different mutations. Hence, the clinical manifestation and severity of mitochondrial disease are highly variable among individuals carrying the same pathogenic mutation. In addition, genetic testing from one tissue is not the representative of heteroplasmy level in other tissues. Hence, a negative result or low level of heteroplasmy in the peripheral blood does not rule out mtDNA-related mitochondrial disease, as the heteroplasmy level may be above threshold in other tissues such as brain and muscle. The heteroplasmy level is not fixed. In rapidly dividing tissues such as blood, the daughter cells may receive a different proportion of mutant mtDNA than the precursor cell due to random segregation during cell division (also called somatic segregation). Cells with a higher proportion of mutation may be at disadvantage and over the time may get overpopulated by cells with lower mutation proportion, thus the overall heteroplasmy level may decline over time.

MITOCHONDRIAL DYNAMICS

Mitochondria are not static, isolated organelles but they form a dynamic interconnected tubular network in cells. Mitochondria constantly undergo fusion and fission to maintain its steady state. Two adjacent mitochondria fuse to form an elongated mitochondrion. Mitochondrial fusion leads to admixture of mitochondrial matrix and thus allows dilution of accumulating mtDNA mutations and matrix proteins damaged by reactive oxygen species (ROS). Thus, it helps to maintain bioenergetic efficiency. Fusion also allows enhanced communication with endoplasmic reticulum necessary for calcium hemostasis. A mitochondrion divides into two or more daughter mitochondria by the process of fission. Fission is necessary for mitochondrial distribution during the cell division. Fission also helps to segregate a damaged portion of mitochondria so that it can be processed for a special form of autophagy called "mitophagy." Mitochondrial fusion and fission play essential roles in mitochondrial bioenergetic efficiency, quality control, and redistribution. The main protein related to mitochondrial fission is dynamin-related protein1 (*DRP1*). It forms a helical constricting structure over mitochondria at the site of fission. Recruitment of dynamin-related protein 1 to the outer surface of mitochondria is mediated by different receptors such as fission 1 homolog protein (*FIS1*) and mitochondrial fission factor (*MFF*). The fusion of mitochondrial outer membrane is mediated by mitofusin 1 and 2 (*MFN1/MFN2*), while the fusion of inner membrane is mediated by the protein, optic atrophy 1 (*OPA1*). The impairment of mitochondrial fusion and fission processes is related to many inherited and acquired diseases.

INHERITANCE OF MITOCHONDRIAL DISEASES

Mitochondrial diseases are genetically heterogeneous. It can be caused by a pathogenic mutation in mtDNA, or mutations in one of the several nuclear genes required for ETC unit formation, their import into the mitochondria, and assembly as complexes as well as mtDNA replication, transcription, and translation. Disorders of abnormal mitochondrial fusion and fission are also caused by nuclear gene mutations. Inheritance of a mitochondrial disease depends on whether it is caused by mtDNA or nuclear gene mutation.

Inheritance of mtDNA-Related Mitochondrial Disease

Mitochondrial diseases can be caused by deletion of a portion of mitochondrial DNA or point mutations. Mitochondrial deletion-related disorders are usually sporadic. Mitochondrial disease caused by pathogenic mtDNA point mutation is maternally inherited as mitochondrial DNA in the zygote is almost exclusively derived from the egg as the sperm contains relatively fewer mitochondria and those are removed before fertilization. An affected mother carrying homoplasmic pathogenic mtDNA mutation will pass this to all of

TABLE 3.2
Inheritance Pattern of Mitochondrial Diseases

Inheritance Pattern	Example
Sporadic	Mitochondrial DNA deletion syndromes
Maternal	Mitochondrial DNA point mutation syndromes
X-linked	Barth syndrome (*TAZ*)
Autosomal dominant	Autosomal dominant optic atrophy (*OPA1*)
Autosomal recessive	Most of the disorders caused by nuclear gene mutations

her offspring who will then develop a mitochondrial disease. However, they may or may not develop a clinical disease if the mother is heteroplasmic for the mutation because the levels of heteroplasmy will vary between the offspring. An unaffected mother with heteroplasmy level below the threshold may give birth to an affected child or an affected mother with heteroplasmy above the threshold may give birth to an unaffected child. This is explained by the "mitochondrial bottleneck" phenomenon.

Mitochondrial Bottleneck
During oogenesis, as primordial germ cell divides to form primary oocytes, there is a rapid reduction in the number of mitochondria per cell. Primary oocyte then undergoes maturation to form mature oocyte that is associated with replication of mitochondria. This reduction followed by replication in mitochondrial number leads to different proportions of mtDNA with mutation in the mature oocyte than the primordial germ cell. Hence, a mutation below the threshold level in the primordial germ cell may reach above the threshold level in the mature oocyte.

Inheritance of Nuclear Gene-Related Mitochondrial Disease
Inheritance of mitochondrial diseases caused by nuclear gene mutations follows the Mendelian pattern. Inheritance can be autosomal recessive, autosomal dominant, or X-linked (Table 3.2).

MITOCHONDRIAL HAPLOGROUPS
Inheritance of mitochondrial DNA is unique in several ways—it is maternally inherited and there is no intermolecular recombination during transmission. Hence, mtDNA of the entire human population can be traced back to the common ancestor mother (Mitochondrial Eve). However, mtDNA is also prone to accumulate changes at a higher rate due to its relatively inefficient repair mechanism and a higher rate of mutation due to its proximity to ROS-generating respiratory chain. A mtDNA variation arises in a single mtDNA molecule, and over several generations may become fixed (homoplasmic). Hence, mtDNA sequence shows variation between indigenous populations of different geographical regions. Mitochondrial haplogroup can be defined as a population who share similar mtDNA sequence (or have similar mtDNA sequence changes or polymorphism). Mitochondrial haplogroups are named with letters A—Z and are of much interest in understanding human history and migration patterns. Different haplogroups are also considered to increase susceptibility to or be protective against certain acquired diseases.

SUGGESTED READING
1. Gorman GS, Chinnery PF, DiMauro S, et al. Mitochondrial diseases. *Nat Rev Dis Primers.* 2016;2:16080.
2. Boczonadi V, Horvath R. Mitochondria: impaired mitochondrial translation in human disease. *Int J Biochem Cell Biol.* 2014;48:77—84.
3. Stewart JB, Chinnery PF. The dynamics of mitochondrial DNA heteroplasmy: implications for human health and disease. *Nat Rev Genet.* 2015;16:530—542.
4. Archer SL. Mitochondrial dynamics—mitochondrial fission and fusion in human diseases. *N Engl J Med.* 2013;369:2236—2251.

CHAPTER 4

Diagnosis of Mitochondrial Diseases

ABSTRACT

Mitochondrial diseases are clinically and genetically heterogeneous. Diagnosis of mitochondrial diseases can be very challenging. There is no established protocol for diagnosis of mitochondrial disease. An understanding of different diagnostic methods and its limitations is key to timely diagnosis. Fortunately, advancement in molecular diagnostic technologies has enabled rapid diagnosis in many circumstances. This chapter outlines commonly utilized diagnostic modalities.

KEYWORDS

Laboratory diagnosis; Mitochondrial disease; Muscle biopsy; Neuroimaging; Next generation sequencing.

BIOCHEMICAL SCREENING

Blood lactate is often elevated (>2.1 mmol/L) in mitochondrial diseases. However, elevated lactate is neither a sensitive nor a specific marker of mitochondrial disease. It can be elevated in conditions such as hypoxia, circulatory shock, or sepsis. On the other hand, a mitochondrial disease may not be associated with high lactate. Table 4.1 outlines the biochemical abnormalities seen in mitochondrial diseases. Although, none of the biochemical abnormalities mentioned is diagnostic, a combination of abnormalities in the absence of an alternative explanation is more suggestive of a mitochondrial disorder. Confirmatory diagnostic workup should be undertaken if biochemical parameters are suggestive or even when biochemical-screening tests are normal but there is very strong clinical suspicion for mitochondrial disease.

MOLECULAR GENETIC TESTS

The molecular genetic test has rapidly evolved to become the first-line investigation for confirmation of a suspected mitochondrial diagnosis. Next-generation sequencing has enabled to sequence a large number of genes simultaneously thus significantly reducing the time and cost to obtain molecular confirmation. Next-generation sequencing has also increased the likelihood of molecular confirmation and discovery of novel disease-causing genes. Mitochondrial gene sequencing can be combined with nuclear gene sequencing further enhancing the efficiency. However, negative mitochondrial gene sequencing from blood does not rule out mitochondrial DNA-related disease because of heteroplasmy. If suspicion of mitochondrial

TABLE 4.1 Biochemical-Screening Tests in Suspected Mitochondrial Disease	
Test	**Remarks**
Blood lactate	Blood lactate is neither a sensitive nor a specific test for mitochondrial disease. Proper collection is necessary to avoid erroneously high value. Normal blood lactate should be followed with specific diagnostic tests when a mitochondrial disease is clinically suspected.
Creatine phosphokinase (CPK), liver enzymes	An elevated CPK and elevated liver enzymes may indicate mitochondrial myopathy and hepatopathy, respectively.
Plasma amino acids	Elevated alanine may suggest a chronic elevation in blood lactate.
Urine organic acids	Urine organic acid analysis may show tricarboxylic acid cycle intermediates such as malate, and fumarate. In addition, it can show elevated lactic acid, ethylmalonic acid, and 3-methylglutaconic acid.
Plasma carnitine	Analysis of plasma total and free carnitine may show low free carnitine.

Mitochondrial Medicine. https://doi.org/10.1016/B978-0-12-817006-9.00004-6

TABLE 4.2
Molecular Genetic Diagnostic Approaches for Mitochondrial Disease

Molecular Genetic Test	Remarks
Targeted mitochondrial DNA genetic testing for point mutation and/or deletion	When the clinical presentation is typical, targeted genetic test allows confirmation of a clinical diagnosis. For example, testing for common mtDNA mutations for MELAS (Mitochondrial Encephalopathy Lactic Acidosis and Stroke-like episodes)
Targeted nuclear gene sequencing	Example: Sequencing of *POLG1* when clinical presentation (childhood onset progressive encephalopathy, seizures, liver failure) is typical.
Targeted nuclear gene panel	When there are findings indicating a particular condition that is genetically heterogeneous. For example, a substantial decrease in mtDNA content in muscle biopsy specimen may be due to mutations in one of the many genes implicated in mitochondrial DNA maintenance. These genes can be sequenced together as mtDNA depletion panel.
Combined mitochondrial genome and nuclear gene panel	A popular approach is to combine mtDNA sequencing with the sequencing of known nuclear genes related with mitochondrial function. The number of nuclear genes varies between different laboratories offering this test and not always comprehensive. Approximately, 1500 nuclear genes are directly or indirectly related with mitochondrial function.
Whole exome/genome sequencing with mitochondrial genome sequencing	This is the most comprehensive approach. It not only allows diagnosis of a mitochondrial disorder but also other disorders that are in the differential diagnosis. However, there is a likelihood of finding more variants of unknown significance.

disease remains high after negative mtDNA sequencing from a blood specimen, mtDNA sequencing should be performed on tissues that are clinically more severely affected. For example, sequencing of mtDNA from muscle biopsy will enable the diagnosis of a mtDNA-related mitochondrial disease when skeletal muscle is involved. Moreover, it will also enable additional studies such as histopathology, enzymatic, and electron microscopic studies on the biopsy specimen. Genetic tests can be targeted to one or more genes when the clinical phenotype is typical for a known mitochondrial disease. A mitochondrial gene panel should be considered when the presentation is not typical for a known mitochondrial disease but highly suspicious for a mitochondrial disorder such as multisystemic presentation with high lactate and other biochemical screening suspicious for a mitochondrial disorder. For a more vague presentation, where differential diagnosis is broad, a whole exome/genome sequencing approach is more logical and time/cost effective. Table 4.2 summarizes the different molecular genetic diagnostic approaches.

INVASIVE TISSUE DIAGNOSIS
An invasive diagnostic approach such as liver or muscle biopsy is undertaken when the molecular genetic studies from blood are unremarkable, and there is

high suspicion for a mitochondrial disorder. Sometimes, mitochondrial studies are performed when a biopsy is obtained for a different purpose. For example, mitochondrial studies may be added to liver biopsy specimen obtained for the evaluation of liver disease if a mitochondrial disease is in differential diagnosis. It is critical to select the affected tissue for maximum diagnostic yield. Skeletal muscle biopsy is the most often performed invasive procedure for the diagnosis of mitochondrial disease. Different studies such as histochemistry, mtDNA content assay, mtDNA sequencing, respiratory chain assay, coenzyme Q10 assay, and electron microscopy can be performed on biopsy specimen. Table 4.3 summarizes the different studies that can be performed on skeletal muscle biopsy specimen but may also apply to other tissues.

NEUROIMAGING
Neurological abnormalities such as seizure, developmental delay, neuroregression, ataxia, and stroke are predominant features of many mitochondrial diseases. Hence, neuroimaging is often obtained as part of diagnostic workup. The neuroimaging findings may be nonspecific but sometimes characteristic pattern indicates a particular diagnosis. Table 4.4 enlists common brain MRI findings seen in mitochondrial diseases.

TABLE 4.3
Common Mitochondrial Studies Performed on Skeletal Muscle Biopsy Specimen

Diagnostic Test	Remarks
Histochemistry	Mitochondrial myopathy is often associated with proliferation of mitochondria in the subsarcolemmal region that can be seen as red granular deposits (**ragged-red fibers**) with modified Gomori trichrome stain. Staining with NADH dehydrogenase, succinate dehydrogenase, and cytochrome *c* oxidase provides information about the activity of complexes I, II, and IV, respectively. A normal succinate dehydrogenase activity in the setting of decreased cytochrome *c* oxidase activity may suggest mitochondrial disease caused by mtDNA mutation as the complex II is entirely coded by the nuclear genome.
Respiratory chain (electron transport chain) enzyme assay	Fresh frozen specimen is needed for this assay. Biopsy specimen is homogenized to study individual complexes in the presence of specific substrates by spectrophotometry. Activities of complex I (NADH dehydrogenase), complex II (succinate dehydrogenase), complex III (cytochrome *c* reductase), and complex IV (cytochrome *c* oxidase) can be performed individually or in combination (complex I + III—NADH: cytochrome *c* reductase; complex II + III—succinate: cytochrome *c* reductase). A single complex defect suggests a defect in the coding region of mtDNA, or nuclear genes which encode that particular complex subunit or a factor involved in the assembly of that particular complex. Multiple complex defects suggest the generalized defect such as mitochondrial translation disorder or other nuclear gene defects with secondary effects on mtDNA, or mtDNA defect involving a tRNA gene. Normal complex I and II activities when studied in isolation but deficient I + III and II + III activities suggests coenzyme Q10 deficiency as coenzyme Q10 accepts electrons from both complexes I and II and passes it over to complex III in the electron transport chain.
Coenzyme Q10 quantification	Deficient coenzyme Q10 in muscle biopsy specimen is suggestive of disorders of coenzyme Q10 biosynthesis.
Mitochondrial DNA content estimation	Mitochondrial DNA content is severely reduced in mitochondrial DNA depletion syndromes caused by mutations in genes for mitochondrial DNA maintenance. An increased mitochondrial DNA content indicates compensatory mitochondrial proliferation in mitochondrial disease.
Mitochondrial DNA sequencing and deletion/duplication analysis	Mitochondrial DNA from a biopsy specimen may show pathogenic changes even when the assay form blood is normal because of heteroplasmy. Multiple deletions in mitochondrial DNA are seen in disorders of mitochondrial DNA maintenance.
Electron Microscopy	Electron microscopy is useful to study the structural abnormalities of mitochondria. Abnormal mitochondrial shape and size, abnormal cristae, and abnormal mitochondrial inclusions are some of the findings seen in mitochondrial diseases.

TABLE 4.4
Common Neuroimaging Findings in Mitochondrial Diseases

Neuroimaging Finding	Remarks
Bilateral lesions in basal ganglia and brain stem	Bilateral symmetrical lesions in basal ganglia and brain stem are seen in the Leigh syndrome. The Leigh syndrome is characterized by onset in infancy with rapid neurological decline and characteristic neurological findings. Mutations in both mitochondrial and nuclear genes have been found in the Leigh syndrome.
Stroke	Stroke in nonvascular territory should raise suspicion of mitochondrial disease. Association of nonvascular stroke, seizures, migraine, cognitive decline, muscle weakness, and lactic acidosis is seen in MELAS.
Cortical atrophy and leukodystrophy	Cortical atrophy and white matter abnormalities are nonspecific findings seen in many mitochondrial diseases.
Cerebellar atrophy	Predominant cerebellar atrophy is seen in *POLG1*-related mitochondrial disease.

Magnetic Resonance Spectroscopy

Although MRI provides structural information of the brain, magnetic resonance spectroscopy provides information about the biochemical composition of the area of interest. Lactate in the brain can be estimated in a noninvasive manner by this tool. High lactate is suggestive (but not diagnostic) of mitochondrial disease. Another metabolite of interest estimated by this method is N-acetyl aspartate (NAA). NAA is synthesized in mitochondria and is a marker of neuronal integrity. Mitochondrial disease is often associated with a decrease in NAA.

SUGGESTED READING

1. Mitochondrial Medicine Society's Committee on Diagnosis, Haas RH, Parikh S, et al. The in-depth evaluation of suspected mitochondrial disease. *Mol Genet Metabol.* 2008;94: 16−37.
2. Wong LJ. Next generation molecular diagnosis of mitochondrial disorders. *Mitochondrion.* 2013;13:379−387.
3. Vincent AE, Ng YS, White K, et al. The spectrum of mitochondrial ultrastructural defects in mitochondrial myopathy. *Sci Rep.* 2016;6:30610.
4. Gropman AL. Neuroimaging in mitochondrial disorders. *Neurotherapeutics.* 2013;10:273−285.

CHAPTER 5

Treatment of Mitochondrial Diseases

ABSTRACT

Treatment of mitochondrial diseases is mainly symptomatic. However, exciting clinical trials are underway and new therapies are being explored. This chapter outlines the principles of treatment of mitochondrial diseases, both current and future.

KEYWORDS

Antioxidants; Coenzyme Q10; Mitochondrial diseases; Mitotoxic; Treatment.

GENERAL THERAPY

This class of therapy applies for most of the mitochondrial disorders. It consists of diet, lifestyle, and avoidance of mitotoxic substances.

Diet in Mitochondrial Disease

There is no specific dietary recommendation for mitochondrial disease. Prolonged fasting should be avoided. Meals should be small but frequent. A heart-healthy, nutritionally balanced diet containing complex carbohydrates as the main source of carbohydrates is recommended. When a mitochondrial disease is associated with intractable epilepsy, the ketogenic diet has been found to be useful. Some patients with mitochondrial disease have swallowing dysfunction. When this is suspected, feeding evaluation and appropriate treatment are needed to optimize nutrition.

Lifestyle

Exercise is beneficial in mitochondrial disease. Endurance exercises trigger mitochondrial proliferation and subsequently ATP generation is enhanced. However, exercise should be supervised and tailored to individual patients.

Avoidance of Mitotoxic Substances

Drugs with known mitotoxic effect should be avoided. Table 5.1 enlists medications known to have mitochondrial toxicity. Apart from these medications, steroid has been shown to be associated with deterioration in Kearns-Sayer syndrome (a disorder of mitochondrial DNA deletion characterized by a triad of age of onset before 20 years, pigmentary retinopathy, and progressive external ophthalmoplegia).

PHARMACOLOGICAL THERAPY

Pharmacological therapies for mitochondrial diseases can be divided into nonspecific (aimed to improve overall mitochondrial function) and specific (aimed toward a specific defect).

Nonspecific Pharmacological Interventions

Treatment of mitochondrial diseases at present mostly consists of supplements that are considered to enhance mitochondrial function. They can be used as single agents in a sequential manner or as a combination (mitochondrial cocktail). Table 5.2 enlists common mitochondrial supplements.

Specific Pharmacological Interventions

Unfortunately, there is a lack of specific pharmacological interventions for mitochondrial diseases at present. For some mitochondrial diagnosis, certain medications have been found to be more helpful than others and should always be considered. Table 5.3 enlists such medications.

SYMPTOMATIC THERAPY

Mitochondrial diseases often present as multisystemic illness. A multidisciplinary approach is needed for optimal management. Symptomatic therapies are tailored for individual patients to treat different associated complications. It can be classified as pharmacological interventions, surgical interventions, and other nonpharmacological interventions.

Pharmacological Interventions

Mitochondrial diseases are often multisystemic. Different categories of medications are needed to treat numerous complications seen in mitochondrial diseases.
Seizure—anticonvulsants
Spasticity—muscle relaxants
Headache—analgesics
Cardiac rhythm disturbances—antiarrhythmics

Mitochondrial Medicine. https://doi.org/10.1016/B978-0-12-817006-9.00005-8

TABLE 5.1
Medications With Known Mitochondrial Toxicity

Category	Medication	Remarks
Anticonvulsants	Valproate	Hepatopathy
Antibiotics	Aminoglycoside (Streptomycin, Gentamycin)	Hearing loss
	Chloramphenicol	Inhibits mitochondrial protein synthesis
	Tetracycline	Inhibits mitochondrial protein synthesis
	Linezolid	Long-term use was found to be associated with lactic acidosis, peripheral neuropathy, optic neuropathy, and myelosuppression.
Antiretroviral	Zidovudine	Causes mitochondrial toxicity by various mechanisms including inhibition of replication of mtDNA. The toxicities observed are neuropathy, myopathy, and hepatopathy
Analgesic/ antiinflammatory	Aspirin	Reye syndrome
	Acetaminophen	Hepatopathy
Anticancer	Platinum chemotherapeutics (Cisplatin)	Cisplatin causes mtDNA-platinum adduct formation, improper mitochondrial protein synthesis, increased ROS generation, and apoptosis
Antidiabetic	Metformin	Lactic acidosis
Antihypertensive	Beta-blocker	Exercise intolerance
Anticholesteremic	Statins	Reduction of muscle coenzyme Q10 may lead to myopathy

TABLE 5.2
Mitochondrial Supplements (Nonspecific Use)

Medication/ Supplement	Dose (Pediatric)	Dose (Adult)	Remarks
Coenzyme Q10 (it is available in the reduced form called ubiquinol and oxidized form ubiquinone. Reduced form has better bioavailability and is preferred)	Ubiquinol 2—8 mg/kg in two divided doses Ubiquinone 10—30 mg/kg in two divided doses	Ubiquinol 60—600 mg daily in two divided doses Ubiquinone 300—2400 mg in 2—3 divided doses	CoQ10 is an integral part of the electron transport chain, CoQ10 supplements are considered to enhance the efficiency of the electron transport chain.
Riboflavin (Vitamin B 2)	50—200 mg daily in 2—3 divided doses	50—400 mg daily in 2—3 divided doses	Riboflavin is a precursor of a flavoprotein, which is one of the building blocks of complexes I and II.
Alpha lipoic acid	25 mg/kg/day	300—600 mg/day	Alpha lipoic acid acts as antioxidant scavenging the toxic ROS formed in excess in mitochondrial diseases.
Vitamin E	1—2 IU/kg/day	100—200 IU daily	Acts as antioxidant scavenging the toxic ROS formed in excess in mitochondrial diseases.
Vitamin C	5 mg/kg daily	50—200 mg daily	Acts as antioxidant scavenging the toxic ROS formed in excess in mitochondrial diseases.

TABLE 5.2
Mitochondrial Supplements (Nonspecific Use)—cont'd

Medication/Supplement	Dose (Pediatric)	Dose (Adult)	Remarks
L-carnitine	25–100 mg/kg/day in 2–3 divided doses	1000–3000 mg per day in 2–3 divided doses	L-carnitine facilitates entry of long chain fatty acid in mitochondria for oxidation and removes toxic acyl compounds. Levels can be monitored in blood and dose adjusted accordingly. At high doses, side effects such as fishy body odor and gastric upset may be seen.
L-creatine	0.1 g/kg daily divided into three doses (maximum daily dose 10 g)	2–10 g daily divided into three doses	Creatine phosphate acts as an intracellular buffer for ATP.

TABLE 5.3
Mitochondrial Supplements (Specific Use)

Disease	Remarks	Medication/Supplement	Dose
Primary Coenzyme Q10 deficiency	Primary coenzyme Q10 deficiency is a multisystemic disease caused by defects in genes responsible for CoQ10 synthesis. CoQ10 in high doses should always be considered as it may reverse some of the complications and can limit disease progression.	Coenzyme Q10	CoQ10 in high doses (5–30 mg/kg/daily in children and up to 2400 mg daily in adults) have been used
Mitochondrial encephalomyopathy lactic acidosis and stroke-like episodes (MELAS)	MELAS is a multisystemic disorder with typical onset in childhood. It manifests with seizures, recurrent vomiting, myopathy, stroke, and lactic acidosis.	L-arginine is useful in the treatment of acute stroke episodes and prophylaxis of stroke. Stroke is considered secondary to impaired vasodilation due to mitochondrial angiopathy. L-arginine is converted to nitric oxide (NO) by endothelial nitric oxide synthase. NO is a vasodilator.	Treatment of acute stroke: • 500 mg/kg as IV bolus within 3 h of the onset of symptoms • 500 mg/kg as continuous 24 h IV infusion for 3–5 days. Prophylaxis of stroke: 100–300 mg/kg/day by mouth in three divided doses.
Leber hereditary optic neuropathy (LHON)	LHON is characterized by bilateral, painless, subacute vision failure in young adults. Treatment is largely symptomatic.	Idebenone has shown to be beneficial in this condition. It is CoQ10 analog with higher efficacy	Recommended dose is 300 mg by mouth three times daily
Ethylmalonic encephalopathy (EE)	EE is an early onset progressive disorder characterized by developmental delay, hypotonia, seizure, and generalized microvascular damage manifesting as relapsing purpura and hemorrhagic diarrhea. There is a deficiency of sulfur dioxygenase needed for detoxification of hydrogen sulfide	N-acetylcysteine (NAC) is a precursor of glutathione that buffers hydrogen sulfide. It is used with metronidazole that limits sulfide production from intestinal bacteria.	100 mg/kg/day by mouth for chronic treatment and intravenous for management of acute encephalopathy have been used.

Hypothyroidism—thyroid supplements
Diabetes—insulin
Anemia—iron, folic acid

Surgical Interventions

Surgical interventions are sometimes needed to treat the complications.
Hearing loss—cochlear implant
Ptosis—corrective surgery
Heart block—pacemaker implantation

Other Interventions

Physical therapy
Occupational therapy
Speech therapy
Artificial ventilation for respiratory failure

ORGAN TRANSPLANTATION

The scope of organ transplantation is currently limited in mitochondrial disorder. However, in some conditions, organ transplantation has been performed. For example, liver transplantation can be considered in deoxyguanosine kinase deficiency when the presentation is isolated liver failure or with minimal neurological involvement.

RESEARCH/EXPERIMENTAL THERAPIES

Although treatment options for a mitochondrial disease are currently limited and unsatisfactory, several treatments are being evaluated. The main experimental therapeutic strategies are outlined here.

Enhancing Mitochondrial Biogenesis

Mitochondrial biogenesis is a physiological response to increased energy demand. Hence, if this process can be enhanced, it will lead to an overall increase in the ATP production and improvement in the clinical status. Mitochondrial biogenesis is mainly mediated by the activation of peroxisome proliferator-activated receptor gamma coactivator 1 α (PGC1α). The PGC1α activation leads to the activation of several transcription factors including nuclear respiratory factors (NRF1 and 2) and peroxisome proliferator-activated receptors (PPARs). These lead to the increased transcription of nuclear genes related to mitochondrial biogenesis and function. The PGC1α activation can be caused by elevated AMP mediated by AMP-activated protein kinase (AMPK), and increased NAD+ mediated by Sirtuin-1. Several agents are being investigated to enhance the mitochondrial proliferation (Table 5.4).

Scavenging of ROS

The reactive oxygen species (ROS) generation is increased in mitochondrial diseases due to inefficient electron transport chain. It is considered to play a major role in the pathogenesis of mitochondrial diseases. The excessive production of ROS from mitochondria leads to the damage of membrane lipids, proteins, enzymes, and DNA. This may ultimately lead to cell death. Several antioxidants such as Vitamins E and C are used in mitochondrial disease to mitigate the effect of excessive ROS. Some new agents considered to have enhanced antioxidant properties are under trial (Table 5.5).

TABLE 5.4
Agents That Enhance Mitochondrial Biogenesis and are Currently Being Evaluated

Drug/Agent	Mechanism of Action	Remarks
Bezafibrate	It is a PPAR agonist and PGC1α activator	A clinical trial in mitochondrial myopathy has recently been completed by New Castle University https://clinicaltrials.gov/ct2/show/NCT02398201
Aminoimidazole Carboxamide Ribonucleoside (AICAR)	Causes PGC1α activation by activation of AMPK	Animal studies have shown promising results
Nicotinamide Riboside (NR)	It is a natural precursor of NAD+ and causes PGC1α activation by activation of Sirtuin-1.	Has only been tried in mouse models.
RTA 408	Activates NRF2	A clinical trial for mitochondrial myopathy is underway. https://clinicaltrials.gov/ct2/show/NCT02255422

TABLE 5.5		
Antioxidant Agents in Clinical Trials for Mitochondrial Diseases		
Drug/Agent	**Mechanism of Action**	**Remarks**
EPI 743	It is a vitamin E analog that has strong antioxidant properties and increases intracellular reduced glutathione level.	Studies are currently underway. Initial results are encouraging.
RP 103	It is cysteamine bitartrate that reacts with cysteine in lysosome forming free cysteine and cysteine-cysteamine. Free cysteine is then transported out to the cytoplasm where it acts as a precursor of glutathione.	A long-term study of RP 103 in mitochondrial diseases has been completed. https://clinicaltrials.gov/ct2/show/NCT02473445

Replenishing Nucleosides/Nucleotides

Some mitochondrial diseases are characterized by the depletion of mitochondrial DNA due to inadequate mitochondrial nucleotides. One such example is Thymidine Kinase 2 (TK2) deficiency. TK2 participates in the mitochondrial pyrimidine salvage pathway by causing phosphorylation of nucleosides deoxycytidine and deoxythymidine to generate deoxycytidine monophosphate and deoxythymidine monophosphate, respectively. These are then converted to their corresponding nucleotides. In the TK2 deficiency due to lack of deoxythymidine and deoxycytidine nucleotides, mitochondrial replication is impaired leading to mtDNA depletion. This disorder is characterized by rapidly progressive myopathy leading to respiratory failure and death. Supplementation with nucleosides deoxycytidine and deoxythymidine as well as deoxycytidine monophosphate and deoxythymidine monophosphate has been shown to be beneficial in mouse models. A similar disorder of the mitochondrial purine salvage pathway is deoxyguanosine kinase deficiency, which causes phosphorylation of purine deoxynucleosides. Similar strategy as in the TK2 deficiency can be used for the therapy.

Inhibition of Permeability Transition Pore

Permeability transition pore (PTP) is a transient channel in the inner mitochondrial membrane that forms at the time of mitochondrial stress such as excessive intramitochondrially ionized calcium, decreased mitochondrial membrane potential (MMP), damage by ROS, and very low ATP levels. The opening of this pore leads to dissipation of MMP, depletion of ATP, osmotic swelling of mitochondria, release of apoptotic factors into the cytoplasm, and ultimately cell death. This mechanism is considered to be important in the pathogenesis of mitochondrial diseases. Hence, agents that can inhibit PTP are being explored. Cyclosporine A is one such agent and is being investigated as a form of treatment in Leber hereditary optic neuropathy (LHON).

Restoring Abnormal Calcium Homeostasis

In many mitochondrial diseases, intramitochondrial calcium concentration is decreased, which contributes to the decreased ATP production. The blockage of mitochondrial Na^+/Ca^{2+} exchanger that extrudes calcium from mitochondria in exchange of sodium has shown to increase intramitochondrial calcium and ATP production. This may be a potential therapeutic option. CGP37157 inhibits mitochondrial Na^+/Ca^{2+} exchanger and is being evaluated in animal models.

Molecular Genetic Approaches

Gene therapy: the approach of gene therapy for mitochondrial diseases caused by nuclear gene defects is very similar to gene therapy for other inherited diseases where a vector containing the gene of interest is targeted to body cells. Once inside the cell, the vector carried gene starts expressing wild-type protein thus offsetting the effect of disease-causing genetic changes. However, there are several challenges with this approach such as the reach of the vector to the whole body and concerns of the vector-induced genetic changes. Adeno-associated virus vectors are being studied in mouse models of myopathy and other mitochondrial diseases caused by nuclear gene mutations. The results from these studies are promising and have the potential to be translated into clinical practice. Gene therapy for mitochondrial DNA-related mitochondrial disease is more challenging because to be effective the gene of interest will have to reach and be expressed in all or most of the mitochondria in a cell. One approach to gene therapy of mtDNA-related mitochondrial disease is to tag the gene of interest with mitochondrial targeting sequence (MTS). Hence after transfection, the gene of interest is expressed in the nucleus but after translation the protein contains MTS, enabling its entry into the mitochondria. One such study for LHON is under clinical trial and recruiting patients (https://clinicaltrials.gov/ct2/show/NCT02161380).

Modulation of heteroplasmy: mtDNA-related mitochondrial diseases manifest clinically only when the mutation level reaches above threshold. Hence, if the heteroplasmy level can be decreased by eliminating or reducing the mtDNAs with mutation, clinical improvement may be expected. Several methods of recognizing specific mtDNA sequence and subsequent digestion by nuclease are currently being studied in cell lines and animal models. These methods include mitochondria-targeted restriction endonucleases, mitochondria-targeted zinc finger nucleases, and mitochondria-targeted transcription activator-like effector nucleases. The principle behind all these methods is the same—recognizing a specific mtDNA sequence (mutation) and subsequent cleavage of mtDNA containing that specific mutation. This leads to a decrease in mtDNAs containing mutation and overpopulation by wild-type mtDNA thus mitigating the effect of mtDNA mutation on a cellular level.

Bypassing the block in respiratory chain: respiratory chain is organized differently in some lower vertebrates, plants, and yeast, For example, in yeast, the enzyme NADH/CoQ reductase substitutes complex 1 by directly passing electron to CoQ. Similarly, in some species, CoQ/O2 alternative oxidase (AOX) provides an alternative route for electron by bypassing complexes III and IV. These proteins can be utilized to bypass the block in respiratory chain caused by nuclear or mtDNA mutations. The NADH/CoQ reductase can be expressed to bypass complex I block while AOX can bypass the block in either complexes III or IV. These alternative enzymes do not propel proton across the inner mitochondrial membrane but do restore the flow of electron and hence ultimately lead to increased ATP production. These approaches are being investigated in cell lines and animal models.

SUGGESTED READING

1. Parikh S, Saneto R, Falk MJ, et al. A modern approach to the treatment of mitochondrial disease. *Curr Treat Options Neurol.* 2009;11:414—430.
2. Viscomi C, Bottani E, Zeviani M. Emerging concepts in the therapy of mitochondrial disease. *Biochim Biophys Acta.* 2015;1847:544—557.
3. El-Hattab AW, Zarante AM, Almannai M, Scaglia F. Therapies for mitochondrial diseases and current clinical trials. *Mol Genet Metabol.* 2017;122:1—9.
4. Schon EA, DiMauro S, Hirano M, Gilkerson RW. Therapeutic prospects for mitochondrial disease. *Trends Mol Med.* 2010;16:268—276.

CHAPTER 6

Management of Emergency and Surgery

ABSTRACT

Patients with mitochondrial disease are prone to deteriorate further during the times of catabolic stressors such as illness, fever, prolonged fasting, and surgery. This chapter outlines the principles of management during sickness and surgery to avoid deterioration of mitochondrial function.

KEYWORDS

Anesthesia; Critical care; Emergency; Mitochondria; Surgery.

MANAGEMENT OF EMERGENCY

The common metabolic emergency in adults and children with mitochondrial disease is an infection, either viral or bacterial. During illness, appetite is decreased and there is increased energy need. This leads to catabolic stress on mitochondria. The baseline symptoms may worsen during such stress and even new symptoms may emerge due to failing mitochondria. Certain measures can be undertaken to avoid or minimize deterioration of mitochondrial function.

1. Avoid fasting. If oral intake is inadequate, consider hospitalization and intravenous fluid therapy.
2. Management of hospitalized patients:
   a. Start dextrose-containing intravenous fluid— Usually, 5% or 10% of dextrose-containing intravenous fluid at 1.25−1.5 times the maintenance rate. **Ringer Lactate is contraindicated.**
   b. Monitor routine chemistries, lactic acid, glucose, ammonia, and liver function.
   c. For worsening lactic acidosis, dextrose concentration in the intravenous fluid may need to be decreased if high dextrose concentration is being used.
   d. For severe metabolic acidosis (pH < 7.2), sodium bicarbonate should be given as a bolus (1 mEq/kg) followed by an infusion. For less severe acidosis, bicarbonate may be added to the intravenous fluid.
   e. Hyperammonemia can be managed by promoting anabolism with high dextrose (10% or higher) containing intravenous fluid. If the patient develops hyperglycemia, an insulin infusion (0.05−0.1 unit/kg/hour) should be started. Very high ammonia (>200 μmol/L) will require ammonia scavenger therapy (sodium benzoate/sodium phenylacetate infusion).
   f. Evaluation and treatment of precipitating illness and comorbid complications should be done simultaneously.
   g. Intravenous levocarnitine at 100 mg/kg/day in four divided doses should be considered. Patient's oral mitochondrial supplements may be stopped and restarted when it is safe to start oral medications.
   h. Avoid any mitotoxic medications (Chapter 5, Table 5.1).
   i. Enteral feeding should be started when the patient is stable and it is safe to start enteral feeds. Intravenous fluid should be tapered and stopped once enteral feeding has resumed and returned to the baseline.

MANAGEMENT OF SURGERY

Surgery is a catabolic stress, and hence careful management of surgery is critical to avoid deterioration of a mitochondrial condition. Following are general principles of perioperative management of a mitochondrial disease patient.

1. Preoperative management—Detailed preoperative evaluation to determine the severity of comorbidities and anticipated complications during surgery is needed. Fasting should be minimized. If possible, the surgery should be scheduled for the first slot in the morning. For minor surgery, a patient may arrive early in the morning when intravenous fluid should be started. For major surgery, a patient may need to be admitted on the previous day. Intravenous fluid should be started when the patient is made NPO. Intravenous fluid should contain dextrose. **Ringer lactate is contraindicated.**
2. Intraoperative management—Intravenous fluid should be continued. Body temperature, heart rhythm, blood glucose, and acid−base status should be closely monitored. Many anesthetic agents have shown to decrease mitochondrial function in vitro.

However, anesthetic agents are tolerated well in vivo. In general, the following should be considered for anesthesia/sedation in patients with mitochondrial disease:

a. Avoid anesthetic agents that are complex1 inhibitors, such as halothane and barbiturates (phenobarbitone).

b. Volatile anesthetics can depress mitochondria. Sevoflurane is better tolerated than halothane, isoflurane, or desflurane. If possible, a lower dose should be used.

c. For sedation, propofol should be avoided. If propofol use cannot be avoided, it should be given as bolus for short procedures. Prolonged propofol infusion should be avoided. Benzodiazepines are safer alternatives for sedation.

d. Muscle relaxants are safe. However, caution should be exercised in patients with preexisting myopathy and respiratory muscle weakness.

3. Postoperative management—Mitochondrial patient needs to be monitored for longer than patients without mitochondrial disease. Before extubation, careful monitoring is needed to ensure that there is no prolonged inhibitory effect from the neuromuscular blocker. Intravenous fluid needs to be continued until normal PO/enteral intake has been established.

SUGGESTED READING

1. Miyamoto Y, Miyashita T, Takaki S, Goto T. Perioperative considerations in adult mitochondrial disease: a case series and a review of 111 cases. *Mitochondrion*. 2016;26: 26−32.
2. Footitt EJ, Sinha MD, Raiman JA, Dhawan A, Moganasundram S, Champion MP. Mitochondrial disorders and general anaesthesia: a case series and review. *Br J Anaesth*. 2008;100:436−441.

Genetic Counseling, Prenatal Diagnosis, and Reproductive Options in Mitochondrial Diseases

ABSTRACT

Mitochondrial diseases are clinically and genetically heterogeneous. Proper counseling, prenatal diagnosis, and reproductive choices for a couple with family history of mitochondrial disease depend heavily on the molecular diagnosis. This chapter outlines the approach and challenges of genetic counseling, prenatal diagnosis, and reproductive options in mitochondrial diseases.

KEYWORDS

Autosomal; Genetic counseling; Inheritance; Maternal inheritance; Nuclear transfer; Prenatal diagnosis.

GENETIC COUNSELING

Genetic counseling depends on whether mitochondrial disease is caused by mutation(s) in mitochondrial DNA or nuclear gene.

Mitochondrial Disease due to Nuclear Gene Mutation

Mitochondrial disease due to nuclear gene mutation can be inherited as autosomal recessive, autosomal dominant, or X-linked manner (Table 7.1).

Autosomal dominant inheritance: A patient may have inherited the mutation from one of the parents who may or may not be symptomatic or it may be de novo in the patient. Parents should be tested and if one of them carries the mutation, the risk of each of the patient's siblings of carrying this mutation is 50% at conception. Each offspring of the patient has a 50% risk of inheriting this mutation.

Autosomal recessive inheritance: The parents of an affected patient are usually carriers and asymptomatic. At conception, each sibling of the patient has a 25% chance of inheriting both mutations from parents and thus being affected, a 50% chance of inheriting one mutation and thus being a carrier, and a 25% chance of inheriting none of the mutations thus being neither a carrier or affected. The offspring of an individual with the autosomal recessive condition are obligate heterozygotes (carriers).

X-linked inheritance: A patient may have inherited the mutation from his mother who is usually asymptomatic or it may be de novo in the patient. If the mother carries the mutation, the risk of transmitting it in each of her pregnancy is 50%. Male siblings of the patient who inherit the mutation will be affected while females will be carriers. A female inheriting the mutation may be affected due to nonrandom inactivation of most of the X chromosomes without mutation (skewed inactivation) or when the condition is X-linked dominant or semidominant (pyruvate dehydrogenase deficiency). All daughters of a male patient will be carriers and none of his sons will be affected.

Mitochondrial Disease due to Mitochondrial DNA Mutation

Mitochondrial disease can be due to mitochondrial DNA deletion or point mutation.

Mitochondrial disease due to mitochondrial DNA deletion: Mitochondrial DNA deletion is usually de novo. The risk to sibs of a patient is very low. If a male is carrying mitochondrial DNA deletion, his

TABLE 7.1 Inheritance Pattern of Mitochondrial Diseases Caused by Nuclear Gene Mutations	
Inheritance Pattern	**Example**
Autosomal dominant	Autosomal dominant optic atrophy (*OPA1*)
Autosomal recessive	Most of the disorders caused by nuclear gene
X-linked	Barth syndrome (*TAZ*)

Mitochondrial Medicine. https://doi.org/10.1016/B978-0-12-817006-9.00007-1

offspring will not inherit the deletion. If a female is affected, there is a small risk (estimated to be about 1 in 24) that she will have an affected child.

Mitochondrial disease due to mitochondrial DNA point mutation: Mitochondrial DNA point mutation is maternally inherited. Mother of a patient carries the mutation but may or may not be symptomatic due to a different level of heteroplasmy than the patient. All the siblings of the patient are at risk of inheriting the mutation from the mother but may or may not be symptomatic due to genetic bottleneck resulting in a different level of heteroplasmy among them (chapter 3). If a male is carrying mitochondrial DNA point mutation, his offspring will not inherit the mutation. If an affected female is heteroplasmic, all her offspring will inherit the mutation, but they may or may not be affected due to genetic bottleneck resulting in a different level of heteroplasmy between her and her offspring and among her offspring.

Mitochondrial Disease due to Unknown Molecular Etiology

Counseling is challenging if the molecular diagnosis is not established. A detailed family history may give a clue to the mode of inheritance, but can be misleading sometimes. A maternal inheritance will imply mitochondrial DNA mutation while the history of consanguinity will suggest the autosomal recessive inheritance. If muscle biopsy shows isolated complex II deficiency, the inheritance is likely autosomal recessive as complex II is entirely nuclear in origin. It is estimated that approximately 75% of adult-onset mitochondrial disease is caused by mitochondrial DNA mutation while it accounts for only approximately 25% of childhood-onset mitochondrial disease.

PRENATAL DIAGNOSIS

Prenatal Diagnosis of Mitochondrial Disease due to Nuclear Gene Mutation

Chorionic villus sampling (CVS): CVS is performed between 10 and 12 weeks of gestational age. Placental tissue is tested for the known mutation. Whether fetus will be affected, carrier or unaffected is determined.

Preimplantation genetic diagnosis (PIGD): It involves in vitro fertilization and genetic testing of cells removed from the embryos at a very early stage. Suitable embryos are then implanted.

Prenatal Diagnosis of Mitochondrial Disease due to Mitochondrial DNA Mutation

Chorionic villus sampling: Placental tissue is tested for heteroplasmy and the risk of the fetus being affected is estimated on this basis.

Preimplantation genetic diagnosis: Heteroplasmy level in cells removed from the embryos at a very early stage is measured. Embryos with heteroplasmy below threshold (18%) are then implanted.

Prenatal diagnosis of mitochondrial disease caused by mitochondrial DNA mutation is challenging due to the concern of variation in heteroplasmy levels between placental tissue and the fetus (CVS) or among cells of early embryo (PIGD). However, recent studies suggest that heteroplasmy levels detected by these methods correlate well with those in the fetus (CVS) and the rest of the embryo (PIGD).

REPRODUCTIVE OPTIONS

Reproductive Options for Couple with Family History of Mitochondrial Disease due to Nuclear Gene

Reproductive option for such couples is prenatal diagnosis and determination of whether fetus may be affected. Sperm donation or oocyte donation may be considered in suitable circumstances. Table 7.2 summarizes feasible options based on the mode of inheritance.

Reproductive Options for a Female Carrier of Pathogenic Mitochondrial DNA Mutation

CVS or PIGD can provide reliable information on heteroplasmy level of fetus/embryo. However, the possibility

TABLE 7.2 Reproductive Options for a Couple with a Family History of Mitochondrial Disease due to the Nuclear Gene	
Inheritance Pattern	**Reproductive Options**
Autosomal dominant	CVS, PIGD, Sperm/oocyte donation (based on who carries the mutation)
Autosomal recessive	CVS, PIGD, Sperm donation, Oocyte donation
X-linked	CVS, PIGD, Oocyte donation

of giving birth to an affected child when low hetero-plasmy level was detected by one of these methods cannot be completely excluded as heteroplasmy level may vary between fetus and placenta (CVS) and between cells of the embryo (PIGD). The interpretation becomes even more complicated for an intermediate level of heteroplasmy found by these methods. More specific information about a particular mutation should be sought by literature review before making a decision. In women with very high level of heteroplasmy or homoplasmic mutation, these methods may consistently show a high load of mutation in placental/embryo cells. In such situations, prevention of transmission of mitochondrial disease can be achieved by in vitro fertilization of donor oocyte with sperm from male partner and implantation of the embryo in the recipient (oocyte donation). Many couples want a genetically related child. For such couples, choices are very limited and only available at few places (UK) or on research/experimental basis. One such option is mitochondrial donation or germline nuclear transfer. This technique involves removal of nuclear genome from an oocyte or zygote carrying mitochondrial DNA mutation followed by transfer to an enucleated donor oocyte or zygote with healthy mitochondria. Thus, the resulting embryo will have healthy mitochondria from the donor but nuclear genetic materials from both parents. Last but not least, adoption is always an option!

SUGGESTED READING

1. Thorburn DR, Dahl HH. Mitochondrial disorders: genetics, counseling, prenatal diagnosis and reproductive options. *Am J Med Genet.* 2001;106:102−114.
2. Nesbitt V, Alston CL, Blakely EL, et al. A national perspective on prenatal testing for mitochondrial disease. *Eur J Hum Genet.* 2014;22:1255−1259.
3. Richardson J, Irving L, Hyslop LA, et al. Concise reviews: assisted reproductive technologies to prevent transmission of mitochondrial DNA disease. *Stem Cell.* 2015;33:639−645.
4. Craven L, Tang MX, Gorman GS, De Sutter P, Heindryckx B. Novel reproductive technologies to prevent mitochondrial disease. *Hum Reprod Update.* 2017;23:501−519.

CHAPTER 8

Mitochondrial Diseases: Classification

ABSTRACT

Mitochondrial genetic material is incomplete. Approximately 1500 proteins encoded by nuclear genes are required for mitochondrial biogenesis and function. Mitochondrial diseases can be broadly classified as due to mitochondrial DNA mutations or nuclear DNA mutations. Those caused by nuclear DNA mutations can further be classified based on whether they affect respiratory chain unit formation, mitochondrial DNA maintenance, transcription, translation, or mitochondrial homeostasis. In addition, tricarboxylic acid cycle and pyruvate metabolism disorders closely resemble mitochondrial diseases and hence often discussed with mitochondrial diseases.

KEYWORDS

Classification; Genetics; Mitochondrial disease; Mitochondrial DNA; Nuclear gene.

INTRODUCTION

Mitochondrial disease can be caused by a mutation in either mitochondrial or nuclear DNA. Mitochondrial genetic material is incomplete. Nuclear genes are required for mitochondrial biogenesis and mitochondrial functions. Approximately 1500 proteins encoded by nuclear genes are required for mitochondrial function. Mitochondrial diseases can be broadly classified as due to mitochondrial DNA mutations and nuclear DNA mutations.

MITOCHONDRIAL DISEASES CAUSED BY MITOCHONDRIAL DNA MUTATIONS

Mitochondrial DNA diseases can be divided into diseases caused by mtDNA rearrangements (such as deletion) and mtDNA point mutations. Mitochondrial DNA point mutations can lead to the impaired synthesis of individual respiratory chain subunit or impairment of mitochondrial protein synthesis machinery (Table 8.1).

MITOCHONDRIAL DISEASES CAUSED BY NUCLEAR GENE MUTATIONS

Approximately 300 nuclear genes have been associated with mitochondrial diseases. Nuclear gene defects can

TABLE 8.1 Mitochondrial Diseases Caused by mtDNA Mutations	
Mitochondrial DNA Rearrangement	**Example**
Mitochondrial DNA deletion	Kearns-Sayre syndrome
	Pearson syndrome
	Chronic progressive external ophthalmoplegia (CPEO)
Mitochondrial DNA Point Mutation	**Example**
Mutations in genes encoding respiratory chain subunit	Leber hereditary optic neuropathy (LHON)
	Neurogenic muscle weakness, ataxia, and retinitis pigmentosa (NARP)
	Maternally inherited Leigh syndrome (MILS)

Continued

Mitochondrial Medicine. https://doi.org/10.1016/B978-0-12-817006-9.00008-3

TABLE 8.1
Mitochondrial Diseases Caused by mtDNA Mutations—cont'd

Mutations impairing mitochondrial protein synthesis	Mutations in mitochondrial tRNA genes	Mitochondrial myopathy, encephalopathy, lactic acidosis, and stroke-like episodes (MELAS)
		Myoclonic epilepsy with ragged red fibers (MERRF)
		Maternally inherited diabetes and deafness (MIDD)
		Hypertrophic cardiomyopathy
	Mutations in mitochondrial ribosomal RNA genes	Aminoglycoside-induced deafness
		Nonsyndromic hearing loss

cause mitochondrial disease by several mechanisms such as impairment of respiratory complex subunit synthesis, respiratory complex assembly, mitochondrial DNA replication, mitochondrial DNA translation, or mitochondrial homeostasis (Table 8.2). In addition, disorders of tricarboxylic acid cycle and pyruvate metabolism are often classified under mitochondrial diseases.

TABLE 8.2
Nuclear Genes Associated With Mitochondrial Disease

Respiratory Complex Structural Subunit	*Gene Examples*
Complex 1	*NDUFV1, NDUFV2, NDUFS1, NDUFS2, NDUFS3, NDUFS4, NDUFS6, NDUFS7, NDUFS8, NDUFA1, NDUFA2, NDUFA10, NDUFA11, NDUFA12,*
Complex II	*SDHA, SDHB, SDHC, SDHD*
Complex III	*UQCRB, UQCRQ, UQCRC2, CYC1*
Complex IV	*COX8A, COX7B, COX6B1, COX6A1, COX4I2*
Complex V	*ATP5A1, ATP5E*
Respiratory Complex Unit Assembly	*Gene Examples*
Complex 1	*NDUFAF1, NDUFAF2, NDUFAF3, NDUFAF4, NDUFAF5, NDUFAF6, ACAD9, FOXRED1, NUBPL*
Complex II	*SDHAF1, SDHAF2*
Complex III	*BCS1L, TTC19, LYRM7, UQCC2, UQCC3*
Complex IV	*SURF1, COA5*
Complex V	*ATPAF2, TMEM70*
Mitochondrial DNA Maintenance	*Gene Examples*
Mitochondrial DNA replication	*POLG, POLG2, TWNK, TFAM, RNASEH1, DNA2, MGME1*
Maintenance of mitochondrial nucleotide pool	*TK2, DGUOK, SUCLA2, SUCLG1, ANT1(SLC25A4), MPV17, TYMP, RRM2B*
Mitochondrial DNA Translation	*Gene Examples*
Mitochondrial ribosomal protein	*MRPL3, MRPL12, MRPL44, MRPS16, MRPS22, MRPS34, MRM2, ERAL1*
Mitochondrial translation factors	*GFM1, TUFM, TSFM, C120RF65, RMND1*
Mitochondrial tRNA modification	*TRNT1, TRMU, MTO1, GTPBP3, NSUN3, TRMT5, TRIT1, PUS1*

TABLE 8.2 Nuclear Genes Associated With Mitochondrial Disease—cont'd	
Mitochondrial aminoacyl tRNA synthetase	*AARS2, CARS2, DARS2,EARS2, FARS2, HARS2, IARS2, LARS2, MARS2, NARS2, PARS2, RARS2, SARS2, TARS2, VARS2,WARS2, YARS2, GARS, KARS, MTFMT*
Mitochondrial Homeostasis	***Gene Examples***
Mitochondrial membrane integrity	*AGK, TAZ, DNAJC19*
Mitochondrial proteostasis	*HSPD1, CLPP, SPG7, AFG3L2*
Mitochondrial dynamics	*OPA1, MFN2, DNM1L*
Pyruvate Metabolism and Tricarboxylic Acid Cycle	***Gene Examples***
Pyruvate dehydrogenase complex	*PDHA1, PDHB, DLAT, DLD, PDHX, PDP1, TPK1, LIAS,*
Pyruvate carboxylase	*PC*
Tricarboxylic acid cycle	*OGDH, SUCLG1, SUCLA2, SDHA, SDHB, SDHC, SDHD, FH*

Mitochondrial DNA Deletion Disorders

ABSTRACT

Mitochondrial diseases due to mitochondrial DNA mutations can be due to mitochondrial DNA deletion or point mutation. This chapter describes the clinical presentation, diagnosis, and principles of management of mitochondrial DNA deletion disorders.

KEYWORDS

Kearns–Sayre syndrome; Mitochondrial DNA deletion; Ophthalmoplegia; Pearson syndrome; Progressive external ophthalmoplegia.

CLINICAL PRESENTATION

Mitochondrial DNA deletion disorders consist of three clinically recognizable overlapping phenotypes—chronic progressive external ophthalmoplegia (CPEO), Kearns-Sayre Syndrome (KSS), and Pearson syndrome (Table 9.1). Although CPEO is characterized by progressive bilateral ptosis and reduction of ocular motility, KSS is a multisystemic disorder with a characteristic triad of age of onset below 20 years of age, pigmentary retinopathy, and progressive external ophthalmoplegia (PEO). In addition to the characteristic triad, one of three other features (heart conduction block, cerebellar ataxia, and cerebrospinal fluid protein concentration greater than 100 mg/dL) is required for the diagnosis of KSS. Often, patients present with PEO and multisystemic manifestation without the characteristic triad of KSS (PEO plus). Pearson syndrome is characterized by sideroblastic anemia and pancreatic insufficiency. Anemia and ptosis are the most common initial presentations. Although these syndromes are classic mitochondrial DNA deletion disorders, the initial presentation can be failure to thrive, renal tubulopathy, or endocrinopathy. Mitochondrial DNA deletion disorders should be considered in any patient with unexplained neurological disorder particularly in the presence of multisystem involvement. A detailed evaluation for systemic involvement is needed to establish the extent of disease. Occasionally, mitochondrial DNA deletion syndrome can manifest as Leigh syndrome characterized by early onset neuroregression, and lesions in basal ganglia and brain stem. Rarely, MELAS (mitochondrial encephalomyopathy, lactic acidosis, and stroke-like episodes), MERRF (myoclonic epilepsy, ragged red fibers), and MIDD (maternally inherited diabetes and deafness) have also been reported with mitochondrial DNA deletions.

GENETICS

Mitochondrial DNA deletion disorders are caused by deletions of varying sizes. However, the size of deletion is the same in an individual. Also, the same deletion can result in Pearson syndrome, PEO, PEO plus, or KSS. In Pearson syndrome, deletion is most abundant in blood, while it is confined to skeletal muscle in PEO. In KSS, deletion is present in all tissues including blood but sometimes may be undetectable in blood. The size of deletion can range from 2 to 10 kb. However, a 4.97 kb deletion (m.8470_13,446del4977) is the most common deletion.

Mitochondrial DNA deletion is usually de novo. They commonly arise by homologous recombination or slipped mispairing during oogenesis or early embryogenesis. The mother of a proband is usually not affected. An affected woman has a very small chance of having an affected child (about 1 in 24). The father of a proband does not have the risk of carrying the same deletion, and the offspring of a male with mtDNA deletion are not at risk.

DIAGNOSIS

Mitochondrial DNA deletion can be detected by the Southern blot technique. Long-range PCR followed by massive parallel sequencing can also detect deletions. Skeletal muscle is the ideal tissue for detection of mtDNA deletion in PEO, PEO plus, and KSS. However, diagnosis of mtDNA deletion disorder can be made from peripheral blood in the pediatric population. Peripheral blood is the preferred tissue for Pearson syndrome.

Mitochondrial Medicine. https://doi.org/10.1016/B978-0-12-817006-9.00009-5

TABLE 9.1
Mitochondrial DNA Deletion Syndromes

Mitochondrial DNA Deletion Syndromes	Main Features
CPEO	Bilateral ptosis, symmetrical reduction of ocular motility, proximal myopathy
KSS	Age of onset <20 years, pigmentary retinopathy, PEO, and one of the three following features: • Heart conduction block • cerebellar ataxia • cerebrospinal fluid protein concentration greater than 100 mg/dL
PEO plus	PEO, multisystemic manifestation
Pearson syndrome	Sideroblastic anemia, exocrine pancreatic insufficiency

TABLE 9.2
Systemic Manifestations of mtDNA Deletion Disorders and Their Management

Organ System	Manifestations	Management
Constitutional	Failure to thrive	Optimize nutrition, gastrostomy tube feeding
Ophthalmology	Ptosis, retinopathy	Ptosis surgery
Audiology	Sensorineural deafness	Cochlear implant, hearing aids
Neurology	Intellectual disability, developmental delays, hypotonia, ataxia, seizures	Anticonvulsant
Musculoskeletal	Proximal myopathy	Physical therapy, occupational therapy
Hematological	Anemia	Transfusions
Endocrinology	Diabetes mellitus, pancreatic insufficiency, adrenal insufficiency, thyroid insufficiency, hypoparathyroidism, growth hormone deficiency	Hormone replacement (pancreatic enzymes, thyroid hormone, etc.)
Cardiovascular	Heart block, cardiomyopathy	Pacemaker
Respiratory	Respiratory muscle weakness	Ventilator support
Gastrointestinal	Dysphagia, gastroparesis	Motility agents
Renal	Tubulopathy, chronic renal failure	Periodic monitoring, renal transplant

A muscle biopsy may show ragged red fibers and cytochrome oxidase negative fibers. In addition, the decreased activity of respiratory chain complexes is found. Blood lactate is often elevated. Magnetic resonance imaging of the brain may show lesions in white matter, basal ganglia, or brain stem.

MANAGEMENT

Diagnosis of mitochondrial DNA deletion disorder should lead to an evaluation for other systemic manifestations. Management is largely symptomatic, avoidance of mitotoxic medications, physical and occupational therapy (Table 9.2). Mitochondrial

TABLE 9.3 Surveillance for Systemic Complications in Mitochondrial Deletion Disorders	
Evaluation	**Recommended Interval**
Growth and development (pediatric patients) Weight (adults)	Every visit
ECG, echocardiography	6–12 monthly
Hearing	12 monthly
Hemoglobin A1C, blood glucose, thyroid	6–12 monthly
Renal function tests	6–12 monthly

supplements such as Coenzyme Q10, L-carnitine, Niacin, and Riboflavin can be tried. Periodic evaluation for cardiac, endocrine, renal, and audiological complications is needed (Table 9.3). A multidisciplinary approach (audiologist, neurologist, ophthalmologist, hematologist, endocrinologist, cardiologist, gastroenterologist, nephrologist, physical therapy, and rehabilitation specialist) is needed for optimal management.

SUGGESTED READING

1. Broomfield A, Sweeney MG, Woodward CE, et al. Paediatric single mitochondrial DNA deletion disorders: an overlapping spectrum of disease. *J Inherit Metab Dis.* 2015;38: 445–457.
2. DiMauro S, Hirano M. Mitochondrial DNA deletion syndromes. In: Adam MP, Ardinger HH, Pagon RA, et al., eds. *GeneReviews® [Internet].* Seattle (WA): University of Washington, Seattle; 2003:1993–2018.

CHAPTER 9.1

Progressive External Ophthalmoplegia

CLINICAL PRESENTATION

Progressive external ophthalmoplegia (PEO) is a clinical syndrome characterized by bilateral ptosis and symmetrical reduction of ocular motility. It is a common finding in mitochondrial DNA deletion syndromes—CPEO and KSS. In CPEO, external ophthalmoplegia is an isolated finding or associated with proximal muscle weakness. However, KSS is a multisystemic disorder with a characteristic triad of age of onset below 20 years of age, pigmentary retinopathy, and PEO. In addition to the characteristic triad, one of three other features (heart conduction block, cerebellar ataxia, and cerebrospinal fluid protein concentration > 100 mg/dL) is required for the diagnosis. Often, patients present with PEO and multisystemic manifestations without the characteristic triad of KSS (PEO plus). PEO is a genetically heterogeneous condition. The genetic etiology can be mtDNA deletion, mtDNA point mutation, or nuclear gene mutation(s). Common features associated with PEO secondary to nuclear gene mutations are neuropathy, psychomotor regression, seizures, ataxia, Parkinsonism, dysarthria, and liver abnormalities. A clinical diagnosis of PEO should lead to a detailed evaluation of systemic complications to establish the extent of disease. PEO can be the predominant manifestation or part of a clinically recognizable syndrome. Syndromes associated with PEO are outlined in Table 9.1.1.

Ptosis and ophthalmoparesis can also be seen in other genetic and acquired conditions such as myotonic dystrophy type 1, oculopharyngeal muscular dystrophy, myasthenia gravis, and thyroid ophthalmopathy.

GENETICS

PEO is sporadic in about half of cases caused by de novo mitochondrial DNA deletions. The remaining cases are caused by either autosomal dominant or recessive nuclear gene mutations, or mtDNA point mutations. The genetic etiology of PEO is summarized in Table 9.1.2.

TABLE 9.1.1
Syndromes Associated with PEO

Mitochondrial DNA Deletion Syndromes	Main Features
CPEO	Bilateral ptosis, symmetrical reduction of ocular motility, proximal myopathy
KSS	Age of onset <20 years, pigmentary retinopathy, PEO, and one of the three following features: • Heart conduction block • cerebellar ataxia • cerebrospinal fluid protein concentration greater than 100 mg/dL
Mitochondrial DNA point mutation syndromes	**Main features**
MELAS (mitochondrial encephalomyopathy, lactic acidosis, and stroke-like episodes)	• Stroke-like episodes, seizures, recurrent headaches, psychomotor regression, myopathy • Usually caused by a point mutation in the *MT-TL1* gene that codes for mitochondrial t-RNA for leucine
Mendelian syndromes	**Main features**
SANDO (sensory ataxia, neuropathy, dysarthria, ophthalmoparesis)	• Sensory ataxia, peripheral neuropathy, seizures, ophthalmoplegia • Caused by recessive *POLG* mutations
MNGIE (mitochondrial neurogastro intestinal encephalopathy)	• Severe gastrointestinal dysmotility, cachexia, neuropathy, ophthalmoparesis, myopathy, leukodystrophy) • Caused by recessive *TYMP* mutations
DOA (dominant optic atrophy)	• Bilateral optic atrophy, sensorineural hearing loss, ataxia, myopathy • Caused by dominant *OPA1* mutation

TABLE 9.1.2
Genetics of PEO

Genetic Change	Inheritance	Example	Remarks
Mitochondrial DNA deletion	Sporadic	Common 4.97 kb deletion (m.8470_13446del4977)	• Most common cause of PEO
Mitochondrial DNA point mutation	Maternal	*MT-TL1* *MT-TK* *MT-TQ* *MT-TA*	• PEO can be a predominant clinical manifestation or associated with a clinically recognizable syndrome such as MELAS.
Nuclear gene mutations	Autosomal dominant	*POLG* *OPA1* *ANT1* *PEO1* *RRM2B*	• Commonly associated with other systemic manifestations such as psychomotor regression, seizures, peripheral neuropathy, and ataxia • Commonly associated with genes related to mitochondrial DNA maintenance. Multiple mtDNA deletions are seen in muscle biopsy.
	Autosomal recessive	*POLG* *TYMP* *TK2*	• Commonly seen as part of complex multisystemic disorder such as SANDO and MNGIE • Commonly associated with genes related to mitochondrial DNA maintenance. Multiple mtDNA deletions or mtDNA depletion is seen in muscle biopsy.

TABLE 9.1.3
Mitochondrial DNA Analysis From a Skeletal Muscle Biopsy

Assay	Diagnostic Utility	Remarks
Southern blot	Deletion	Finding of multiple mtDNA deletions is suggestive of an autosomal dominant or recessive condition
Long-range PCR followed by massive parallel sequencing	Point mutation, deletion, heteroplasmy	
Quantitative PCR	Mitochondrial DNA depletion	Finding of mtDNA depletion is suggestive of an autosomal recessive condition

DIAGNOSIS

Skeletal muscle is the ideal tissue for mtDNA analysis in PEO. Table 9.1.3 summarizes the common findings from muscle biopsy mtDNA analysis and its diagnostic importance.

A muscle biopsy may show ragged red fibers and cytochrome oxidase negative fibers. In addition, decreased activity of respiratory chain complexes may be found.

When the presentation is typical of a known disorder such as MNGIE, diagnosis can be made by a targeted genetic test. Combined approach consisting of the sequencing of mtDNA with a panel of nuclear genes related to mitochondrial disease from peripheral blood DNA is often the first-tier genetic test.

MANAGEMENT

A diagnosis of PEO should lead to an evaluation for other systemic manifestations. Management is largely symptomatic, avoidance of mitotoxic medications,

physical and occupational therapy. Mitochondrial supplements such as Coenzyme Q10, L-carnitine, Niacin, and Riboflavin can be tried. For certain diagnosis, more specific therapy is available either as clinical or research basis (L-arginine for MELAS, bone marrow transplantation for MNGIE, nucleoside replacement therapy for TK2). Periodic evaluations for cardiac, endocrine, renal, and audiological complications are needed. A multidisciplinary approach (audiologist, neurologist, ophthalmologist, hematologist, endocrinologist, cardiologist, gastroenterologist, nephrologist, physical therapy, and rehabilitation specialist) is needed for optimal management.

SUGGESTED READING

1. McClelland C, Manousakis G, Lee MS. Progressive external ophthalmoplegia. *Curr Neurol Neurosci Rep.* 2016;16:53.
2. Orsucci D, Angelini C, Bertini E, et al. Revisiting mitochondrial ocular myopathies: a study from the Italian Network. *J Neurol.* 2017;264:1777–1784.

Mitochondrial DNA Point Mutation Disorders

ABSTRACT

Disorders caused by mitochondrial DNA point mutations vary in clinical presentation. They may present as a well-recognized clinical syndrome such as MELAS or symptoms confined to a single organ system such as nonsyndromic hearing loss. Often, the presentation is multisystemic. This chapter outlines the manifestation and management of mitochondrial diseases caused by mtDNA point mutations.

KEYWORDS

LHON; MELAS; MERRF; Mitochondrial DNA; Point mutation; Sensorineural hearing loss.

CLINICAL PRESENTATION

Mitochondrial DNA point mutation disorders present with a wide spectrum of the clinical phenotype. The symptoms can be confined to one organ system (such as nonsyndromic hearing loss and aminoglycoside-induced deafness caused by a mutation in *MT-RNR1*) or multisystemic. Table 10.1 outlines the recognizable clinical phenotypes associated with mtDNA point mutations.

GENETICS

Mitochondrial DNA point mutation disorders are caused by various mutations in genes for mitochondrial messenger RNA, t-RNA, or ribosomal RNA (Table 10.2).

A similar clinical phenotype can be seen with point mutations in different genes (MELAS can be due to *MT-TL1* or *MT-ND1* mutations) and mutation in one gene can cause different phenotypes (*MT-ATP6* mutations can result in NARP or MILS phenotype). The variable presentation of mitochondrial mutations is because of heteroplasmy, tissue distribution of mutation, and the threshold effect.

Mitochondrial disease caused by a mtDNA point mutation is maternally inherited as mitochondrial DNA in the zygote is almost exclusively derived from the egg as the sperm contains relatively fewer mitochondria and those are removed before fertilization. However, the mother of a patient may or may not have symptoms because of a different level of mutation in her than the patient (heteroplasmy). An affected female will pass the mutation to all her offspring who may or may not develop the clinical disease if she is heteroplasmic for the mutation because the levels of heteroplasmy may vary between her and her offspring (bottleneck effect). The maternal siblings of a proband have typically inherited the same mutation from the mother but may be unaffected or have more severe or less severe manifestation (heteroplasmy and bottleneck effect). Occasionally, the point mutation is de novo in which case no other family member of the proband is at risk but a female proband may transmit this de novo mutation to her offspring.

DIAGNOSIS

Mitochondrial DNA point mutation can be diagnosed by targeted genetic testing when the clinical picture is characteristic of a condition such as aminoglycoside-induced ototoxicity. More commonly, a panel of known mutations or mitochondrial genome sequencing is requested. Long-range PCR followed by massive parallel sequencing can detect mtDNA point mutation, deletions, and the heteroplasmy level. In the pediatric population, when the clinical picture is not characteristic of a particular condition, a combined panel approach (mitochondrial genome sequencing with nuclear gene panel) is appropriate as most of the mitochondrial disorders in the pediatric population are due to nuclear gene mutations. Skeletal muscle is an ideal tissue for diagnosis of mtDNA point disorders particularly when there is myopathy. However, many such disorders such as MELAS can be

Mitochondrial Medicine. https://doi.org/10.1016/B978-0-12-817006-9.00010-1

TABLE 10.1
Mitochondrial DNA Point Mutation Syndromes

Mitochondrial DNA Point Mutation Syndromes	Gene	Main Features
MELAS (mitochondrial encephalomyopathy, lactic acidosis, stroke-like episodes)	*MT-TL1* *MT-ND5*	• Stroke-like episodes, seizures, recurrent headaches, psychomotor regression, myopathy • Most commonly caused by m.3243 A > G point mutation in *MT-TL1* gene that codes for mitochondrial t-RNA for leucine
MERRF (myoclonic epilepsy, ragged-red fibers)	*MT-TK* *MT-TF* *MT-TL1* *MT-TI* *MT-TP*	• Myoclonus, generalized epilepsy, ataxia, psychomotor regression, peripheral neuropathy, myopathy, ragged-red fibers on muscle biopsy • Most commonly caused by m.8344 A > G point mutation in *MT-TK* gene that codes for mitochondrial t-RNA for lysine
MIDD (maternally inherited diabetes and deafness)	*MT-TL1*	• Diabetes mellitus, sensorineural hearing loss, pigmentary retinopathy, myopathy, ataxia, cardiomyopathy, ragged-red fiber on muscle biopsy • Most commonly caused by m.3243 A > G point mutation in *MT-TL1* gene that codes for mitochondrial t-RNA for leucine
LHON (Leber hereditary optic neuropathy)	*MT-ND1* *MT-ND4* *MT-ND6*	• Bilateral, painless, subacute visual failure in young adulthood is the most characteristic presentation • Most commonly caused by point mutations in genes for subunits of complex I: m.3460G > A in *MT-ND1*, m.11778G > A in *MT-ND4*, or m.14484T > C in *MT-ND6*
NARP (neurogenic muscle weakness, ataxia, retinitis pigmentosa)	*MT-ATP6*	• Muscle weakness, neuropathy, ataxia, seizures, retinitis pigmentosa • Caused by m.8993T > G (rarely m.8993T > C) point mutation in *MT-ATP6* gene that codes for a complex V subunit.
MILS (maternally inherited Leigh syndrome)	*MT-ATP6* *M-ND3* *MT-ND5* *MT-ND6*	• Early onset neuroregression, hypotonia, ataxia, abnormal movements, breathing irregularity, neuropathy, characteristic bilateral lesions in basal ganglia and brain stem • Caused by m.8993T > G mutation in *MT-ATP6* gene in about half of the cases.

diagnosed from peripheral blood mitochondrial DNA. Diagnosis of mtDNA point mutation disorders can be made from peripheral blood in the pediatric population.

A muscle biopsy may show ragged-red fibers and cytochrome oxidase-negative fibers (MERRF). In MELAS, muscle biopsy may show ragged-red but cytochrome oxidase-positive fibers. In addition, the decreased activity of respiratory chain complexes can be found. Blood lactate is often elevated. Magnetic resonance imaging (MRI) of the brain may show lesions in white matter, basal ganglia, or brain stem. MELAS is characterized by hyperintense lesions mainly in posterior cortex not confined to any vascular area during the stroke-like episodes.

MANAGEMENT

Diagnosis of mitochondrial DNA point disorder should lead to an evaluation for other systemic manifestations. Management is mainly symptomatic, avoidance of mitotoxic medications, physical and occupational therapy (Table 10.3). Aminoglycoside antibiotic is contraindicated when aminoglycoside-induced ototoxicity is

TABLE 10.2 Mitochondrial Genes and Associated Phenotypes			
Mitochondrial DNA Point Mutation	**Gene**		**Associated Clinical Phenotypes**
Mutations in genes for respiratory chain subunit (messenger RNA)	Complex 1 subunits	*MT-ND1*	LHON, MELAS
		MT-ND2	MILS, LHON
		MT-ND3	MILS
		MT-ND4	LHON, MILS, exercise intolerance
		MT-ND5	MILS, MELAS, MELAS/MILS
		MT-ND6	LHON, MILS
	Complex III subunit	*MT-CYB*	Exercise intolerance, hypertrophic cardiomyopathy, multisystemic manifestations
	Complex IV subunits	*MT-CO1*	Deafness, encephalomyopathy, multisystemic disorder
		MT-CO2	Exercise intolerance, rhabdomyolysis, liver failure, multisystemic disorder
		MT-CO3	Myopathy, rhabdomyolysis, encephalopathy, Leigh-like condition
	Complex V subunit	*MT-ATP6*	NARP, MILS
Mutations impairing mitochondrial protein synthesis	Mutations in mitochondrial tRNA genes	Phenylalanine *MT-TF*	Myopathy, MELAS, MERRF
		Valine *MT-TV*	Multisystemic disorder
		Leucine (UUR) *MT-TL1*	MELAS, sensorineural hearing loss, MILS, Diabetes mellitus
		Leucine (CUN) *MT-TL2*	Myopathy, PEO
		Isoleucine *MT-TI*	PEO, cardiomyopathy
		Glutamine *MT-TQ*	Myopathy, encephalomyopathy
		Glutamic acid *MT-TE*	Myopathy, PEO, hypertrophic cardiomyopathy, encephalomyopathy, hearing loss
		Methionine *MT-TM*	Myopathy
		Tryptophan *MT-TW*	Myopathy, encephalomyopathy, MNGIE-like syndrome, MILS
		Alanine *MT-TA*	Myopathy
		Asparagine *MT-TN*	Myopathy, PEO
		Cysteine *MT-TC*	MELAS, multisystemic
		Tyrosine *MT-TY*	Myopathy, PEO
		Serine (UCN) *MT-TS1*	Hearing loss
		Serine (AGY) *MT-TS2*	Retinitis pigmentosa, hearing loss, MERRF/MELAS overlap syndrome
		Aspartic acid *MT-TD*	Myopathy
		Lysine *MT-TK*	MERRF
		Glycine *MT-TG*	Encephalomyopathy, hypertrophic cardiomyopathy

Continued

TABLE 10.2
Mitochondrial Genes and Associated Phenotypes—cont'd

Mitochondrial DNA Point Mutation	Gene		Associated Clinical Phenotypes
	Arginine	*MT-TR*	Encephalomyopathy
	Histidine	*MT-TH*	Hearing loss, MERRF/MELAS overlap syndrome
	Threonine	*MT-TT*	Susceptibility to Parkinson disease
	Proline	*MT-TP*	Myopathy, PEO
Mutations in mitochondrial ribosomal RNA genes	*MT-RNR1*		Aminoglycoside-induced deafness Nonsyndromic hearing loss

TABLE 10.3
Systemic Manifestations of mtDNA Point Mutation Disorders and Their Management

Organ System	Manifestations	Management
Constitutional	Failure to thrive	Optimize nutrition, gastrostomy tube feeding
Ophthalmology	Ptosis, retinopathy, optic neuropathy	Ptosis surgery
Audiology	Sensorineural deafness	Cochlear implant, hearing aids
Neurology	Intellectual disability, developmental delays, hypotonia, ataxia, seizures	Anticonvulsant
Musculoskeletal	Myopathy	Physical therapy, occupational therapy
Endocrinology	Diabetes mellitus, thyroid insufficiency,	Diet, insulin, thyroid replacement, etc.
Cardiovascular	Cardiomyopathy	Angiotensin-converting enzyme inhibitors
Respiratory	Respiratory muscle weakness	Ventilator support
Gastrointestinal	Dysphagia, gastroparesis	Motility agents

suspected. Mitochondrial supplements such as Coenzyme Q10, L-carnitine, Niacin, and Riboflavin can be tried. For certain diagnosis, more specific therapy is available as clinical or research basis (L-arginine for MELAS, Idebenone for LHON). Periodic evaluations for cardiac, endocrine, and audiological complications are needed. A multidisciplinary approach (audiologist, neurologist, ophthalmologist, endocrinologist, cardiologist, gastroenterologist, physical therapy, and rehabilitation specialist) is needed for optimal management.

SUGGESTED READING

1. Wong LJ. Pathogenic mitochondrial DNA mutations in protein-coding genes. *Muscle Nerve.* 2007;36:279—293.
2. Scaglia F, Wong LJ. Human mitochondrial transfer RNAs: role of pathogenic mutation in disease. *Muscle Nerve.* 2008;37:150—171.
3. DiMauro S, Hirano M. MERRF. In: Adam MP, Ardinger HH, Pagon RA, et al., eds. *GeneReviews® [Internet].* Seattle (WA): University of Washington, Seattle; 2003:1993—2018.
4. Naing A, Kenchaiah M, Krishnan B, et al. Maternally inherited diabetes and deafness (MIDD): diagnosis and management. *J Diabet Complicat.* 2014;28:542—546.

CHAPTER 10.1

Mitochondrial Encephalomyopathy Lactic Acidosis and Stroke-Like Episodes

CLINICAL PRESENTATION

Mitochondrial encephalomyopathy lactic acidosis and stroke-like syndrome (MELAS) is characterized by the following:
1. Stroke-like episodes before the age of 40 years
2. Encephalopathy—seizures, psychomotor regression
3. Myopathy—lactic acidosis and/or ragged-red fibers on muscle biopsy

In addition, one of the following is required for diagnosis:
1. Normal early psychomotor development

2. Recurrent vomiting
3. Recurrent headache

However, MELAS is a multisystemic disorder with varying clinical presentation. The stroke-like episode is the clinical hallmark of this condition. It can manifest clinically as focal neurological deficit (motor weakness, cortical visual loss, aphasia), altered sensorium, headache, or seizures. Neuroimaging shows lesions not conforming to a typical vascular territory; hence these episodes are called "stroke-like." Clinical features of MELAS are summarized in Table 10.1.1.

TABLE 10.1.1
Systemic Manifestations in MELAS and Its Management

Organ System	Manifestations	Management
Constitutional	Short stature, failure to thrive	Optimize nutrition, gastrostomy tube feeding
Neurology	Stroke-like episodes, dementia, seizures, myoclonus, hemiparesis, cortical blindness, altered mental status, migraine, ataxia, polyneuropathy, developmental delays, hypotonia, learning disability, basal ganglia classifications, increased cerebrospinal fluid lactate	Anticonvulsants, L-arginine therapy for treatment and prophylaxis of stroke-like episodes
Psychiatric	Depression, anxiety, psychotic disorders	Psychiatric evaluation
Ophthalmology	Optic atrophy, pigmentary retinopathy, progressive external ophthalmoplegia (PEO)	Ptosis surgery
Audiology	Sensorineural deafness	Hearing aid, cochlear implant
Musculoskeletal	Myopathy, exercise intolerance, ragged-red fibers on muscle biopsy, lactic acidosis	Physical therapy, occupational therapy
Endocrinology	Diabetes mellitus, thyroid insufficiency, hypoparathyroidism, growth hormone deficiency, hypogonadism	Diet, oral hypoglycemic agents (avoid metformin), and insulin therapy for diabetes mellitus, thyroid supplements for hypothyroidism
Cardiovascular	Cardiomyopathy, cardiac conduction abnormalities	Pacemaker, angiotensin-converting enzyme inhibitors
Respiratory	Respiratory muscle weakness, pulmonary hypertension	Ventilator support
Gastrointestinal	Recurrent vomiting, constipation, gastroparesis, diarrhea, gastric dysmotility	Motility agents
Renal	Renal Fanconi syndrome, proteinuria, focal segmental glomerulosclerosis	Periodic evaluation, renal transplantation

GENETICS

A point mutation in *MT-TL1* is the most common genetic etiology of MELAS. Mutations in other mitochondrial genes can also cause MELAS. Occasionally, mutations in *POLG* can cause stroke-like episodes and MELAS-like clinical presentation (Table 10.1.2).

MELAS is inherited in maternal fashion. However, mother, siblings, and other maternal relatives of a patient may or may not have symptoms because of different levels of mutation (heteroplasmy) and different tissue distributions of the mutation. The m.3243A > G mutation in *MT-TL1* can cause different phenotypes based on mutation load and tissue distribution (Table 10.1.3).

TABLE 10.1.2
Genetics of MELAS

Genetic Change	Inheritance	Example	Remarks
Mitochondrial DNA point mutation	Maternal	*MT-TL1* *MT-ND5* *MT-TL2* *MT-ND1* *MT-ND5* *MT-TF* *MT-TH* *MT-TK* *MT-TQ* *MT-TS11* *MT-TS2* *MT-TV* *MT-CO2* *MT-CO3* *MT-CYB*	• *MT-TL1* is the most common gene associated with MELAS • Point mutation m.3243A > G is found in 80% of patients with MELAS • Other common *MT-TL1* mutations associated with MELAS are m.3271T > C and m.3252A > G • *MT-ND5* is the second most common gene associated with MELAS. • m.13513G > A is the most common mutation in *MT-ND5* associated with MELAS
Nuclear gene mutation	Autosomal recessive	*POLG*	• *POLG* mutations mimicking MELAS is very rare.

TABLE 10.1.3
Clinical Spectrum of m.3243A > G Mutation

Phenotype	Remarks
Mitochondrial encephalomyopathy lactic acidosis and stroke-like syndrome (MELAS)	Most severe phenotype
Maternally inherited Diabetes mellitus and Deafness (MIDD)	About 85% cases of MIDD are due to m.3243A > G mutation. Diabetes mellitus or deafness can be isolated manifestation at presentation
Progressive external ophthalmoplegia (PEO)	Results from higher mutation load in skeletal muscles
Hypertrophic cardiomyopathy	Cardiomyopathy can be an isolated finding or associated with other systemic manifestations such as diabetes mellitus and hearing loss
Leigh syndrome	Leigh syndrome is genetically heterogeneous and can be due to mutations in the nuclear or mitochondrial gene.
Focal segmental glomerulosclerosis (FSGS)	Proteinuria or renal failure due to FSGS can be the main clinical manifestation or associated with other systemic manifestations.

DIAGNOSIS

Diagnosis of MELAS can be made by a targeted genetic test for common mutation(s) or mitochondrial genome sequencing. Diagnosis can be made from DNA obtained from peripheral blood leukocytes. However, due to varying heteroplasmy level among different tissues, the mutation may be absent or present at very low level in peripheral blood. For confirmation of diagnosis, mutation assay on other tissues such as muscle, urinary epithelial cells, or buccal mucosa is needed in such a scenario.

A muscle biopsy may show ragged-red fibers but cytochrome oxidase-positive fibers. In addition, decreased activity of respiratory chain complexes can be found on muscle biopsy. MRI of the brain shows ischemic lesions not conforming to a vascular territory during the stroke-like episodes. It may also show basal ganglia calcification, brain atrophy, and other sequelae from previous episodes.

MANAGEMENT

A diagnosis of MELAS should lead to an evaluation for other systemic manifestations. Management is mainly symptomatic, avoidance of mitotoxic medications, physical and occupational therapy (Table 10.1.1). Mitochondrial supplements such as Coenzyme Q10, L-carnitine, creatine, and lipoic acid can be tried. Periodic evaluations for cardiac, endocrine, renal, and audiological complications are needed. A multidisciplinary approach (audiologist, neurologist, ophthalmologist, endocrinologist, cardiologist, nephrologist, physical therapy, and rehabilitation specialist) is needed for optimal management. Stroke-like episodes in MELAS is a medical emergency. Early treatment with L-arginine can reduce the severity of stroke-like episodes, while prophylactic use of L-arginine can decrease the frequency of such episodes. Table 10.1.4 outlines the management of stroke-like episodes in MELAS.

TABLE 10.1.4
Management of Stroke-like Episodes in MELAS

Phase of Treatment	Principles of Treatment
Acute: Stroke-like episode can manifest as acute neurological symptoms such as loss of consciousness, visual loss, aphasia, or focal motor deficit.	1. Patient should be stabilized hemodynamically: management of airway, breathing, and circulation. 2. Intravenous dextrose containing fluid should be started: for example, 10% Dextrose with ½ normal saline 3. L-arginine: intravenous L-arginine hydrochloride bolus at the dose of 500 mg/kg mixed with 25 mL/kg of water/D5W/D10W should be given over 90 min. 4. Ideally, a central line should be established for arginine infusion 5. An EEG and brain MRI should be obtained. 6. Vital signs should be monitored for hypotension induced by arginine bolus 7. Following lab should be obtained—blood gas, lactate, blood glucose, routine chemistries 8. After initial bolus: intravenous arginine should be given as a continuous infusion over 24 h at the dose of 500 mg/kg for 3–5 days. If the patient has stabilized and able to take enteral medications, arginine can be given orally at the same daily dose divided 6 hourly. 9. Vital signs, blood glucose, and routine chemistries should be monitored during the infusion 10. Arginine infusion can cause hypotension, acidosis, hyperglycemia, and hyperkalemia that should be monitored and treated accordingly.
Chronic: chronic or prophylactic treatment is aimed to decrease the frequency of stroke-like episodes	L-arginine at the dose of 100–300 mg/kg/day orally in three divided doses.

SUGGESTED READING

1. El-Hattab AW, Adesina AM, Jones J, Scaglia F. MELAS syndrome: clinical manifestations, pathogenesis, and treatment options. *Mol Genet Metabol.* 2015;116:4−12.
2. Koenig MK, Emrick L, Karaa A, et al. Recommendations for the management of strokelike episodes in patients with mitochondrial encephalomyopathy, lactic acidosis, and strokelike episodes. *JAMA Neurol.* 2016;73:591−594.
3. DiMauro S, Hirano M. MELAS. In: Adam MP, Ardinger HH, Pagon RA, et al., eds. *GeneReviews® [Internet].* Seattle (WA): University of Washington, Seattle; 2001:1993−2018.

CHAPTER 10.2

Leber Hereditary Optic Neuropathy

CLINICAL PRESENTATION

Leber hereditary optic neuropathy (LHON) is characterized by painless, subacute, central vision loss. The onset of symptoms is typically in the second to third decade of life. The onset of symptoms is very rare after the age of 50 years. Males are more commonly affected as about 80%−90% of patients with LHON are male.

The clinical course of LHON can be divided into active and atrophic phase. The active phase is characterized by progressive vision loss. There is blurring of vision affecting central visual field at the onset. In most cases, visual acuity progressively declines to 20/200 or worse. The vision loss is usually unilateral at the onset but involves the other eye weeks to months later (median delay 6−8 weeks). In about 25% of cases, vision loss is bilateral at the onset. During the active phase, ocular fundus examination may show optic disc swelling and hyperemia, telangiectasia around the optic disc, vascular tortuosity, and swelling of nerve fiber layer around the optic disc. After the active phase (approximately 6 weeks), the optic disc becomes atrophic. The vision loss does not progress after the active phase is over. However, there may be partial recovery of vision. Earlier age of onset (<20 years), a slowly progressive course, larger optic disc, and m.14484T > C mutation are associated with better recovery.

Occasionally, patients with LHON may have other systemic features such as cardiac arrhythmia, peripheral neuropathy, myopathy, dystonia, or other movement disorders. There is also a known association between LHON mutations and multiple sclerosis like illness, particularly in females of European origin.

GENETICS

Point mutations in three mitochondrial genes for structural subunits of complex I account for about 95% cases of LHON (Table 10.2.1). Apart from these common mutations, other mutations in *MT-ND1*, *MT-ND4*, *MT-ND6*, and mutations in *MT-ND2*, *MT-ND4L*, and

TABLE 10.2.1
Genetics of LHON

Gene	Mutation	Remarks
MT-ND4	m.11778G > A	• Most common mutation • Accounts for about 70% of all cases of LHON • Worst prognosis for partial visual recovery. Partial visual recovery rate is about 4%−25%.
MT-ND6	m.14484T > C	• Most common mutation in French Canadians • Accounts for about 90% of all cases in French Canadians • Best prognosis for partial visual recovery. Partial visual recovery rate is about 37%−58%.
MT-ND1	m.3460G > A	• Accounts for about 13% of all cases of LHON • Intermediate prognosis for partial visual recovery. Partial visual recovery rate is about 20%.

MT-ND5 can cause LHON. Mutations in *MT-ATP6*, *MT-CO3*, and *MT-CYB* have also been reported in occasional individuals or families.

LHON mutations are usually homoplasmic. However, in about 10%–20% of cases heteroplasmy is found. Similar to other mitochondrial point mutations, LHON mutations are inherited in maternal fashion. Mother of a proband usually has the same mutation, and all the siblings of a proband may have inherited the same mutation from their mother. However, LHON has incomplete penetrance. The family history of LHON is negative in about 40% of cases. The incomplete penetrance in LHON is not well understood, and several factors are considered to influence the expressivity of LHON mutation:

1. Gender: Approximately 50% of male and 10% female with LHON mutation will experience vision loss. An X-linked susceptibility factor is considered to modify the disease risk. In addition, estrogen can decrease the severity of mitochondrial dysfunction and mitigate the effect of LHON mutation.
2. Heteroplasmy: It is considered that homoplasmic individuals are more likely to have clinical disease compared to those with heteroplasmy below 60%.
3. Mitochondrial haplogroup: LHON mutations can be more harmful in the context of certain mitochondrial haplogroups. For example, haplogroup J is associated with enhanced penetrance of 11778G > A and 14484 T > C mutations while haplogroup K can increase the penetrance of 3460 G > A mutation.

4. Environmental factors: Heavy smoking is associated with increased penetrance of LHON mutation. Heavy alcohol intake is also considered to increase the penetrance but it is not convincingly proven.

DIAGNOSIS

Diagnosis of LHON can be made by a targeted genetic test for the three common mutations. If there is no mutation detected, mitochondrial genome sequencing should be considered. Diagnosis can be made from DNA obtained from peripheral blood leukocytes.

Visual evoked potential will show evidence of optic neuropathy. Magnetic resonance imaging (MRI) of the brain is usually normal but may show high signal in optic nerves.

MANAGEMENT

Management of LHON is largely symptomatic. It consists of avoidance of smoking, heavy drinking, and mitotoxic medications. Management of optic neuropathy involves low vision aids and vision rehabilitation. Intraocular pressure should be closely monitored and treated. Mitochondrial supplements such as coenzyme Q10, L-carnitine, alpha lipoic acid, and creatine can be tried but there is no evidence of benefit from these. The patient should be evaluated for other systematic manifestations such as ECG for cardiac arrhythmia. Few therapies are currently available for LHON on research basis (Table 10.2.2).

TABLE 10.2.2
Research Therapies for LHON

Type of Treatment	Remarks
Ubiquinone analogs	1. Idebenone is currently available as part of the clinical trial in the United States. The typical dose is 900 mg/day https://clinicaltrials.gov/ct2/show/NCT02774005. 2. EPI-743 has been used previously in a clinical trial with encouraging results.
Gene therapy	1. Gene therapy is currently underway for LHON patients with 11778G > A mutation. https://clinicaltrials.gov/ct2/show/NCT02161380. 2. Genetic engineered AAV virus containing wild-type *MT-ND4* gene tagged with the mitochondria-targeting sequence is used. 3. *MT-ND4* is expressed in the nucleus of the transfected cells but the ND4 protein is imported in mitochondria by the attached mitochondria-targeting sequence.
Stem cells	1. Stem cells can protect retinal ganglion cells by secreting neurotrophic and antiinflammatory factors. 2. Intravitreal injection of stem cells is being studied https://clinicaltrials.gov/ct2/show/NCT03011541.

SUGGESTED READING

1. Yu-Wai-Man P, Chinnery PF. Leber hereditary optic neuropathy. In: Adam MP, Ardinger HH, Pagon RA, et al., eds. *GeneReviews® [Internet]*. Seattle (WA): University of Washington, Seattle; 2000:1993–2018.

2. Meyerson C, Van Stavern G, McClelland C. Leber hereditary optic neuropathy: current perspectives. *Clin Ophthalmol.* 2015;9:1165–1176.

3. DeBusk A, Moster ML. Gene therapy in optic nerve disease. *Curr Opin Ophthalmol.* 2018;29:234–238.

CHAPTER 10.3

Nonsyndromic Hearing Loss: Mitochondrial

CLINICAL PRESENTATION

Sensorineural hearing loss (SNHL) is a common medical condition. The etiology of SNHL may be genetic or environmental. Prelingual SNHL is most commonly genetic (~80%). Most of the genetic SNHL is nonsyndromic (~80%). Nonsyndromic genetic SNHL is usually due to autosomal recessive mutations (~80%). However, about 1% of nonsyndromic genetic SNHL is due to mitochondrial DNA point mutations. The contribution of mitochondrial etiology is higher in postlingual nonsyndromic SNHL where it may account for up to 5%–10% of cases.

Hearing loss is a common manifestation of mitochondrial disorder but usually it is part of a multisystemic disorder such as MELAS and MERRF. Nonsyndromic mitochondrial deafness is characterized by hearing loss of varying severity and onset in the absence of systemic findings. Family history may be suggestive of maternal inheritance. Nonsyndromic mitochondrial SNHL may be insidious in onset or precipitated by aminoglycoside administration (Table 10.3.1).

GENETICS

Mutations in *MT-RNR1* encoding 12s ribosomal RNA and *MT-TS1* encoding mitochondrial transfer RNA for serine are most commonly associated with mitochondrial nonsyndromic SNHL(Table 10.3.2).

The 3243A > G mutation in *MT-TL1* associated with MELAS and MIDD often present with isolated sensorineural hearing loss. Many individuals with this mutation develop diabetes mellitus on follow-up. Other mitochondrial DNA mutations associated with

---

TABLE 10.3.1
Mitochondrial Nonsyndromic Sensorineural Hearing Loss

Clinical Presentation	Remarks
Aminoglycoside-induced ototoxicity	Bilateral, irreversible, severe to profound hearing loss is precipitated within days to weeks of aminoglycoside antibiotic administration. The hearing loss can be precipitated by even a single dose of aminoglycoside and it is different from the dose-dependent hearing loss associated with aminoglycoside use.
Sensorineural hearing loss	Sensorineural hearing loss associated with mitochondrial DNA mutation can present as mild to profound hearing loss of both prelingual and postlingual onset. The usual presentation is bilateral, progressive, postlingual hearing loss. High frequency is preferentially affected in individuals with mild to moderate hearing loss.

TABLE 10.3.2
Genetics of Mitochondrial Nonsyndromic SNHL

Clinical Presentation	Common Genetic Change	Less Common Genetic Changes	Remarks
Aminoglycoside-induced ototoxicity	*MT-RNR1:* 1555A > G	*MT-RNR1:* 1494C > T 961delT/insC *MT-CO1/TS1* boundary: 7444G > A	• The bactericidal effect of aminoglycosides is due to inhibition of bacterial protein synthesis by binding to ribosomal RNA. Mitochondrial protein synthesis is also affected by these antibiotics because of its evolutionary connection to bacteria. *MT-RNR1* mutations make mitochondrial protein synthesis even more susceptible to inhibition by aminoglycosides. • 1555A > G mutation associated with aminoglycoside susceptibility is usually homoplasmic
Sensorineural hearing loss	*MT-RNR1:* 1555A > G *MT-TS1:* 7445A > G	*MT-RNR1:* 1494C > T 961delT/insC 961T > G *MT-TS1:* 7505T > C 7510T > C 7511T > C 7472insC	• *MT-RNR1* 1555A > G is the most common mitochondrial mutation associated with nonsyndromic hearing loss. • *MT-TS1* 7445A > G mutation can sometimes be associated with palmoplantar keratoderma.

nonsyndromic SNHL are as follows: 5655T > C (*MT-TA*), 4295 A > G (*MT-TI*), 12201 T > C (*MT-TH*), and 8296 A > G (*MT-TK*).

Mitochondrial DNA mutations associated with nonsyndromic SNHL have low penetrance. Mitochondrial haplogroup background, other mitochondrial DNA polymorphisms, and nuclear modifier genes are considered to influence the penetrance.

DIAGNOSIS

Diagnosis of nonsyndromic mitochondrial SNHL can be made by a targeted genetic test for common mutations or mitochondrial genome sequencing.

MANAGEMENT

An audiological evaluation should be completed. Depending upon patient's age and preference, the available options are—hearing aid, cochlear implant, sign language training, speech therapy, etc. Mitotoxic medications should be avoided. Special precaution should be taken to avoid aminoglycoside antibiotics in individuals known to have aminoglycoside susceptibility mutation.

SUGGESTED READING

1. Kokotas H, Petersen MB, Willems PJ. Mitochondrial deafness. *Clin Genet.* 2007;71:379−391.
2. Ding Y, Leng J, Fan F, Xia B, Xu P. The role of mitochondrial DNA mutations in hearing loss. *Biochem Genet.* 2013;518: 588−602.
3. Yano T, Nishio SY, Usami S, Deafness Gene Study Consortium. Frequency of mitochondrial mutations in non-syndromic hearing loss as well as possibly responsible variants found by whole mitochondrial genome screening. *J Hum Genet.* 2014;59:100−106.
4. Usami S, Nishio S. Nonsyndromic hearing loss and deafness, mitochondrial. In: Adam MP, Ardinger HH, Pagon RA, et al., eds. *GeneReviews® [Internet].* Seattle (WA): University of Washington, Seattle; 2004:1993−2018.

Mitochondrial Disease of Nuclear Origin: Respiratory Chain Complex Units

ABSTRACT

Mitochondrial respiratory chain (electron transport chain) is located in the inner mitochondrial membrane. It consists of five complexes—complexes I–V. Each complex is made of subunits. Except for complex II, each complex is made up of subunits encoded by both nuclear and mitochondrial genes. Complex II is exclusively nuclear in origin. Mitochondrial diseases can be caused by mutations in nuclear genes that code for respiratory chain complex subunits or its assembly factors. The clinical spectrum is very wide.

KEYWORDS

Assembly factor; Electron transport chain, Respiratory complex unit; Leigh syndrome; Mitochondrial disease.

BACKGROUND

Mitochondrial respiratory chain (electron transport chain) is located in the inner mitochondrial membrane. It consists of five complexes—complexes I–V. Each complex is made of subunits. Except for complex II, each complex is made up of subunits encoded by both nuclear and mitochondrial genes (Table 11.1).

Apart from structural subunits, several assembly factors are required for biogenesis, assembly, and maintenance of complexes. Hence, mitochondrial disease can be a result of mutations in nuclear gene encoding either the structural subunit or assembly factor.

CLINICAL PRESENTATION

Complex I: Complex I deficiency is associated with a wide spectrum of clinical manifestations. It can present with Leigh syndrome, myopathy, hypertrophic cardiomyopathy, lactic acidosis, ataxia, myoclonus, seizures, leukoencephalopathy, ophthalmoplegia, retinopathy, optic atrophy, and failure to thrive. Mutations in both structural subunits of complex I and its assembly factors have been associated with mitochondrial disease (Table 11.2).

Complex II: Complex II is a heterotetramer consisting of four subunits SDHA, SDHB, SDHC, and SDHD. Complex II assembly factors are SDHAF1 and SDHAF2. Complex II is encoded entirely by nuclear genes. Mutations in both genes for structural subunits and assembly factors for complex II are associated with disease (Table 11.3). Mutation in *SDHA* is associated with Leigh syndrome while *SDHAF1* mutations cause infantile leukoencephalopathy. Mutations in *SDHA*, *SDHB*, *SDHC*, *SDHD*, and *SDHAF2* present with paraganglioma. Although Leigh syndrome and infantile leukoencephalopathy caused by complex II genes are autosomal recessive conditions, paraganglioma due to complex II gene mutations is inherited in autosomal dominant fashion.

Complex III: Complex III deficiency is associated with a wide spectrum of clinical manifestations. Mutations in both structural subunits of complex III and assembly factors have been associated with mitochondrial disease (Table 11.4).

Complex IV: Complex IV deficiency is associated with clinical manifestations ranging from encephalomyopathy, liver disease, cardiomyopathy to Leigh syndrome. Mutations in both structural subunits of complex IV and ancillary factors required for biogenesis and assembly of the complex as well as expression of mitochondrial DNA encoded subunits can cause complex IV deficiency (Table 11.5).

Complex V: Complex V deficiency is associated with clinical manifestations ranging from encephalopathy,

Mitochondrial Medicine. https://doi.org/10.1016/B978-0-12-817006-9.00011-3

TABLE 11.1
Genetic Origin of Electron Transport Chain Complexes

Complex	Subunits	Nuclear Encoded	mtDNA Encoded
I	44	37	7
II	4	4	0
III	11	10	1
IV	14	11	3
V	19	17	2

TABLE 11.2
Complex I Deficiency Caused by Nuclear Genes

Mechanism of Deficiency	Gene
Complex I structural subunits	NDUFV1, NDUFV2, NDUFS1, NDUFS2, NDUFS3, NDUFS4, NDUFS6, NDUFS7, NDUFS8, NDUFA1, NDUFA2, NDUFA10, NDUFA11, NDUFA12
Complex I assembly factors	NDUFAF1, NDUFAF2, NDUFAF3, NDUFAF4, NDUFAF5, NDUFAF6, ACAD9, FOXRED1, NUBPL

TABLE 11.3
Complex II Deficiency Genes

Mechanism of Deficiency	Gene
Complex II structural subunits	SDHA, SDHB, SDHC, SDHD
Complex II assembly factors	SDHAF1, SDHAF2

hypertrophic cardiomyopathy, facial dysmorphism, cataract, intrauterine growth retardation, to 3-methylglutaconic aciduria (3-MGA). Mutations in both structural subunits and assembly factors can cause complex V deficiency (Table 11.6).

Coenzyme Q10: CoQ10 is an essential component of the respiratory chain. It accepts electron form complexes I and II and donates to complex III. Hence, deficiency of CoQ10 can cause electron transport chain (ETC) impairment. Moreover, CoQ10 also has antioxidant and antiapoptotic properties. CoQ10 deficiency is classified as primary and secondary. Primary CoQ10 deficiency is due to genetic defects in CoQ10 biogenesis. Secondary CoQ10 deficiency is seen in a large number of genetic and acquired health conditions including many mitochondrial myopathies. Primary CoQ10 deficiency was associated with five distinct clinical phenotypes: encephalomyopathy, cerebellar ataxia, severe infantile multisystemic disease, steroid-resistant nephrotic syndrome, and myopathy. However, the spectrum of clinical presentation of primary CoQ10 deficiency is now recognized to be broader and not confined to these five entities (Table 11.7).

Iron—Sulfur (Fe—S) clusters: Fe—S clusters are essential components of ETC as they perform the crucial function of electron transfer. ETC complexes I, II, and III contain F—S clusters. Apart from electron transfer in the ETC, Fe—S clusters also perform many other vital functions in cells such as control of gene expression, DNA replication, DNA repair, and iron storage. Defects in Fe—S cluster biogenesis are associated with mitochondrial ETC complex deficiency. Table 11.8 outlines the mitochondrial diseases due to F—S cluster biogenesis defect.

Fe—S clusters mainly exist in [2Fe—2S] and [4Fe—4S] forms. Biosynthesis of Fe—S clusters consists of three steps.

Step 1—Fe—S cluster is first assembled on a scaffold protein ISCU. Iron is imported from cytosol across the mitochondrial membrane. Sulfur is provided by the conversion of cysteine to alanine by the desulfurase complex, NFS1-ISD11. [2Fe—2S] clusters are thus formed on the scaffold protein.

Step 2—[2Fe—2S] clusters are released from the scaffold protein by chaperone system and transferred to target proteins by GLRX5.

Step 3—[2Fe—2S] clusters are converted to [4Fe—4S] and targeted to various apoproteins by ISCA1, ISCA2, IBA57, NFU1, and IND1 proteins.

GENETICS

Mitochondrial diseases due to isolated respiratory chain complex deficiency caused by nuclear gene mutations are usually inherited in an autosomal recessive fashion. However, some conditions are inherited in autosomal dominant or X-linked manner (Table 11.9). Occasionally, the condition is sporadic due to de novo mutation.

TABLE 11.4
Complex III Deficiency Caused by Nuclear Genes and Associated Clinical Phenotype

Mechanism of Deficiency	Gene	Clinical Presentation
Complex III structural subunits	UQCRB	Recurrent hypoglycemia, hepatomegaly
	UQCRQ	Psychomotor regression, ataxia, dystonia
	UQCRC2	Recurrent hypoglycemia, hyperammonemia, lactic acidosis
	CYC1	Recurrent hypo/hyperglycemia, reversible neuroregression
Complex III assembly factors	BCS1L	1. Most severe presentation: GRACILE syndrome (Fetal growth **r**etardation, **a**minoaciduria, **c**holestasis, **i**ron overload, **l**actic acidosis, **e**arly death) 2. Intermediate presentation: encephalopathy, liver failure, iron overload, proximal tubulopathy, Leigh syndrome 3. Least severe presentation: Bjornstad syndrome (pili torti, sensorineural hearing loss).
	TTC19	Psychomotor regression, Leigh syndrome
	LYRM7	Psychomotor regression, anemia
	UQCC2	Metabolic acidosis, dysmorphism, psychomotor delay, sensorineural hearing loss
	UQCC3	Lactic acidosis, hypoglycemia, psychomotor delay,

TABLE 11.5
Complex IV Deficiency Caused by Nuclear Genes and Associated Clinical Phenotype

Mechanism of Deficiency		Gene	Clinical Presentation
Complex IV structural subunits		COX8A	Developmental delays, hypotonia, seizures, microcephaly, scoliosis, pulmonary hypertension
		COX7B	Liner skin defect, multiple congenital anomalies
		COX6B1	Encephalomyopathy, leukodystrophy
		COX6A1	Charcot Marie Tooth disease
		COX4I2	Exocrine pancreatic deficiency, dyseryhropoietic anemia
Complex IV ancillary factors	Assembly of early intermediates	SURF1	Leigh syndrome, hypertrichosis, renal tubulopathy, leukodystrophy
		COA5	Fatal neonatal cardioencephalomyopathy
	Copper metalation of mitochondrial DNA encoded subunits (CO1 and CO2) required for complex IV catalytic function	SCO1	Encephalomyopathy, hypertrophic cardiomyopathy, hepatomegaly
		SCO2	Fatal neonatal cardioencephalomyopathy
	Heme incorporation into mitochondrial DNA encoded subunit 1(CO1) required for complex IV catalytic function	COX10 COX15	Leigh syndrome, Leigh-like syndrome, hypertrophic cardiomyopathy
	Maturation of mitochondrial cytochrome oxidase subunits mRNA	LRPPRC	Leigh syndrome – French-Canadian type
	Translational activator of mitochondrial mRNA for CO1	TACO1	Leigh syndrome

TABLE 11.6
Complex V Deficiency Caused by Nuclear Genes and Associated Clinical Phenotype

Mechanism of Deficiency	Gene	Clinical Presentation
Complex V structural subunits	ATP5A1	Neonatal encephalopathy
	ATP5E	Psychomotor delays, peripheral neuropathy, lactic acidosis, 3-MGA.
Complex V assembly factors	ATPAF2	Neonatal encephalopathy, facial dysmorphism, 3-MGA, lactic acidosis
	TMEM70	Neonatal encephalopathy, hypotonia, facial dysmorphism, hypertrophic cardiomyopathy, 3-MGA, lactic acidosis

TABLE 11.7
Genetic Defects of CoQ10 Biosynthesis and Associated Clinical Phenotype

Gene	Clinical Presentation
PDSS1	Optic atrophy, hearing loss, encephalopathy, neuropathy
PDSS2	Retinopathy, hearing loss, Leigh syndrome, ataxia, steroid-resistant nephrotic syndrome
COQ2	Steroid-resistant nephrotic syndrome, encephalopathy, myopathy, MELAS like presentation, retinopathy, hypertrophic cardiomyopathy, neonatal severe multisystemic disease
COQ4	Encephalopathy, seizure, hypotonia, hypertrophic cardiomyopathy
COQ6	Steroid-resistant nephrotic syndrome, sensorineural hearing loss, encephalopathy, seizure
COQ7	Intrauterine growth retardation, persistent pulmonary hypertension of newborn, intellectual disability, hypotonia, polyneuropathy, sensorineural hearing loss
COQ8A	Encephalopathy, ataxia, hypotonia, exercise intolerance
COQ8B	Steroid-resistant nephrotic syndrome
COQ9	Encephalopathy, myopathy, renal tubulopathy, hypertrophic cardiomyopathy

TABLE 11.8
Genetic Defects of Fe–S Cluster Biogenesis and Associated Mitochondrial Diseases

Mechanism of Deficiency	Gene	Clinical Presentation
Assembly of Fe–S clusters on the scaffold protein	ISCU	Hereditary myopathy and lactic acidosis
	NFS1	Infantile complex II/III deficiency (hypotonia, multi organ failure, lactic acidosis)
	ISD11 (LYRM4)	Combined oxidative phosphorylation defect (neonatal lactic acidosis, hypotonia, respiratory failure)
	FXN	Friedreich's ataxia. FXN codes for frataxin which is involved in early steps of F–S cluster assembly
Assembly of [4Fe–4S] and targeting to apoprotiens	NFU1	Multiple mitochondrial dysfunction syndrome type 1: onset soon after birth, hypotonia, neuroregression, respiratory failure
	BOLA3	Multiple mitochondrial dysfunction syndrome type 2: onset in infancy, hypotonia, neuroregression, seizures, leukodystrophy, respiratory failure
	IBA57	Multiple mitochondrial dysfunction syndrome type 3: onset in utero, hypotonia, neuroregression, seizures, brain malformations, leukodystrophy, respiratory failure

DIAGNOSIS

Diagnosis can be made by targeted genetic testing if the clinical picture is characteristic. Different molecular genetic diagnostic approaches are summarized in Table 11.10.

Blood lactate and cerebrospinal fluid lactate are often elevated. Magnetic resonance imaging (MRI) of the brain may show lesions in white matter, basal ganglia, or brain stem. Bilateral lesions involving basal ganglia and brain stem are seen in Leigh syndrome. Magnetic resonance spectroscopy of brain parenchymal tissue may show lactate peak.

Diagnosis of CoQ10 deficiency may be suggested by ETC enzyme assay from a muscle biopsy. Normal complex I and II activities when studied in isolation but deficient I + III and II + III activities suggests CoQ10 deficiency as CoQ10 accepts electron from both complexes I and II and donates it to complex III in the ETC. CoQ10 deficiency can also be diagnosed by quantification in biopsy specimens. Plasma CoQ10 level is misleading as it is affected by diet.

MANAGEMENT

Diagnosis of mitochondrial complex deficiency should lead to an evaluation for other systemic manifestations associated with the deficiency. Management is mainly symptomatic, avoidance of mitotoxic medications, physical and occupational therapy (Table 11.11). Mitochondrial supplements such as Coenzyme Q10, L-carnitine, Niacin, and Riboflavin can be tried. For research trails, clinicaltrials.gov is a very useful resource. Periodic evaluations for cardiac, endocrine, ophthalmologic, and audiological complications are needed. A multidisciplinary approach (audiologist, neurologist, ophthalmologist, endocrinologist, cardiologist, gastroenterologist, physical therapy, and rehabilitation specialist) is needed on an individual case basis for optimal management.

In primary CoQ10 deficiency, early treatment with CoQ10 is beneficial as often the treatment can stop the progression of disease or even reverse some of the manifestations. A high dose of CoQ10 (up to 50 mg/kg/day) is recommended in these conditions.

TABLE 11.9
Inheritance Pattern of Respiratory Chain Complex Nuclear Genes

Inheritance Pattern	Gene	Clinical Phenotype
X-linked	NDUFA1	Complex 1 deficiency Leigh syndrome, retinitis pigmentosa
Autosomal dominant	SDHA SDHB SDHC SDHD SDHAF2	Paraganglioma
Autosomal recessive	SDHA	Leigh syndrome
	Most of the other genes	See text

TABLE 11.10
Molecular Genetic Diagnostic Approaches

Molecular Genetic Test	Remarks
Targeted nuclear gene sequencing	TMEM70: neonatal encephalopathy, hypotonia, facial dysmorphism, hypertrophic cardiomyopathy, 3-MGA, lactic acidosis, Roma ethnicity SURF1: Leigh syndrome, hypertrichosis
Targeted gene panel	If there is isolated complex deficiency on muscle biopsy, panel of genes associated with that complex can be ordered first.
Combined mitochondrial genome and nuclear gene panel	A popular approach is to combine mtDNA sequencing with sequencing of known nuclear genes related with mitochondrial function. If family history is suggestive of autosomal inheritance, only nuclear gene panel can be ordered first.
Whole exome/genome sequencing with mitochondrial genome sequencing	This is the most comprehensive approach. It not only allows diagnosis of a mitochondrial disorder but also other disorder which are in the differential diagnosis. However, there is also increased likelihood of finding variants of unknown significance.

TABLE 11.11
Systemic Manifestations of Mitochondrial Complex Deficiency Diseases and Their Management

Organ System	Manifestations	Management
Constitutional	Failure to thrive	Optimize nutrition, gastrostomy tube feeding
Ophthalmology	Ptosis, retinopathy, optic neuropathy	Ptosis surgery, visual rehabilitation, correction glasses
Audiology	Sensorineural deafness	Cochlear implant, hearing aids, sign language
Neurology	Intellectual disability, developmental delays, hypotonia, ataxia, seizures	Anticonvulsant
Musculoskeletal	Myopathy	Physical therapy, occupational therapy
Endocrinology	Diabetes mellitus, thyroid insufficiency,	Diet, insulin, thyroid replacement, etc.
Cardiovascular	Cardiomyopathy	Angiotensin converting enzyme inhibitors
Respiratory	Respiratory muscle weakness	Ventilator support
Gastrointestinal	Dysphagia, gastroparesis	Motility agents
Renal	Nephrotic syndrome	Angiotensin converting enzyme inhibitors, high dose of CoQ10 in primary CoQ10 deficiency, renal replacement therapy

SUGGESTED READING

1. Fassone E, Rahman S. Complex I deficiency: clinical features, biochemistry and molecular genetics. *J Med Genet.* 2012;49:578–590.
2. Jain-Ghai S, Cameron JM, Al Maawali A, et al. Complex II deficiency–a case report and review of the literature. *Am J Med Genet.* 2013;161A:285–294.
3. Fernández-Vizarra E, Zeviani M. Nuclear gene mutations as the cause of mitochondrial complex III deficiency. *Front Genet.* 2015;6:134.
4. Rak M, Bénit P, Chrétien D, et al. Mitochondrial cytochrome c oxidase deficiency. *Clin Sci.* 2016;130: 393–407.
5. Hejzlarová K, Mráček T, Vrbacký M, et al. Nuclear genetic defects of mitochondrial ATP synthase. *Physiol Res.* 2014; 63:S57–S71.
6. Salviati L, Trevisson E, Doimo M, et al. Primary coenzyme Q10 deficiency. In: Adam MP, Ardinger HH, Pagon RA, et al., eds. *GeneReviews® [Internet]*. Seattle (WA): University of Washington, Seattle; January 26, 2017: 1993–2018.
7. Desbats MA, Lunardi G, Doimo M, Trevisson E, Salviati L. Genetic bases and clinical manifestations of coenzyme Q10 (CoQ 10) deficiency. *J Inherit Metab Dis.* 2015;38: 145–156.
8. Stehling O, Wilbrecht C, Lill R. Mitochondrial iron-sulfur protein biogenesis and human disease. *Biochimie.* 2014; 100:61–77.
9. Cardenas-Rodriguez M, Chatzi A, Tokatlidis K. Iron-sulfur clusters: from metals through mitochondria biogenesis to disease. *J Biol Inorg Chem.* 2018;23:509–520.
10. Xu W, Barrientos T, Andrews NC. Iron and copper in mitochondrial diseases. *Cell Metabol.* 2013;17:319–328.

CHAPTER 11.1

Leigh Syndrome

CLINICAL PRESENTATION

Leigh syndrome (LS) is clinically and genetically heterogeneous condition. The usual presentation is neuroregression precipitated by illness in a child below 2 years of age. There are associated signs and symptoms of basal ganglia and brain stem involvement such as dystonia, ataxia, and abnormal breathing. The condition is progressive and often results in death by 3 years

TABLE 11.1.1
Genetics of Leigh Syndrome

Genome	Biochemical Deficiency		Genes
Nuclear	Isolated respiratory chain complex deficiency	Complex I deficiency	NDUFV1, NDUFV2, NDUFS1, NDUFS2, NDUFS3, NDUFS4, NDUFS7, NDUFS8, NDUFA1, NDUFA2, NDUFA9, NDUFA10, NDUFA12, NDUFAF2, NDUFAF5, NDUFAF6, FOXRED1
		Complex II deficiency	SDHA, SDHAF1
		Complex III deficiency	UQCRQ, BCS1L, TTC19
		Complex IV deficiency	NDUFA4, SURF1, COX10, COX15, SCO2, LRPPRC, TACO1
	Combined OXPHOS deficiency	Mitochondrial DNA replication and maintenance defect	POLG, SUCLA2, SUCLG1, FBXL4
		Mitochondrial DNA translation defect	MTFMT, EARS2, FARS2, IARS2, NARS2, TRMU
		Coenzyme Q10 deficiency	PDSS2
	Pyruvate dehydrogenase complex deficiency		PDHA1, PDHB, PDHX, DLD, SLC19A3, SLC25A19
	Biotinidase deficiency		BTD
Mitochondrial	Isolated respiratory chain complex deficiency	Complex I deficiency	MTND1, MTND2, MTND3, MTND4, MTND5, MTND6
		Complex IV deficiency	MTCO3
		Complex V deficiency	MTATP6
	Combined OXPHOS deficiency	Mitochondrial transfer RNA genes	MTTL1, MTTK, MTTI, MTTV, MTTW
		Mitochondrial DNA deletion	Mitochondrial DNA deletions ranging from 3.6 to 6 kb in size have been reported

of age. There is characteristic bilaterally symmetrical lesion in brain stem and basal ganglia. These lesions show gliosis, vacuolization and vascular proliferation on histopathology and are seen as hyper intense regions in T2-weighted MRI of brain. Other findings sometimes associated with LS are — retinitis pigmentosa, optic atrophy, hypertrophic cardiomyopathy, hepatomegaly, and renal tubulopathy. Often the child's development is normal before the onset of illness or shows only mild delays. Late onset LS and slower progression are also described.

GENETICS

Leigh syndrome is genetically heterogeneous. Mutations in mitochondrial DNA as well as nuclear genes can cause LS. More than 75 genes have been implicated

in the etiology of LS (Table 11.1.1). Mutations in nuclear genes comprise about 80% of LS cases while mitochondrial DNA mutations about 20%. The inheritance pattern of LS can be sporadic, maternal, X-linked, autosomal dominant, and autosomal recessive (Table 11.1.2).

Complex I deficiency is the most common biochemical abnormality in LS. NDUFS4 is the most common nuclear gene while MTATP6 is the most common mitochondrial gene associated with LS.

DIAGNOSIS

Diagnosis of LS is suggested by the characteristic clinical and neuroradiological features. In presence of characteristic clinical findings or inheritance pattern, targeted gene sequencing or mitochondrial genome

TABLE 11.1.2
Inheritance Pattern of Leigh Syndrome

Inheritance Pattern	Gene
Sporadic	Mitochondrial DNA deletion
Maternal	Mitochondrial DNA point mutations
X-linked	PDHA1, NDUFA1
Autosomal dominant	DNM1L
Autosomal recessive	Most of the other genes

TABLE 11.1.3
Etiological Clues for Leigh Syndrome

Clinical Finding	First-Tier Genetic Testing
Maternal inheritance	Mitochondrial genome sequencing or mtDNA point mutations panel
X-linked inheritance, agenesis of corpus callosum, lactate: pyruvate ratio <20	PDHA1
Hypertrichosis	SURF1
Methylmalonic aciduria	SUCLA2, SUCLG1

sequencing can be done first (Table 11.1.3). Muscle biopsy may show isolated respiratory chain complex deficiency in which case genes related with that particular complex can be sequenced first. However, often diagnosis is made by combined panel approach or whole exome sequencing may be needed.

MANAGEMENT

A diagnosis of LS should lead to evaluation for other systemic manifestations. Management is mainly symptomatic, avoidance of mitotoxic medications, physical and occupational therapy (Table 11.1.4). Mitochondrial supplements such as coenzyme Q10, L-carnitine, and riboflavin can be tried. For certain diagnosis, more specific therapy is available (Table 11.1.5); hence a molecular diagnosis should always be attempted when possible. Many research based therapies are becoming available (Table 11.1.6). The comprehensive list of ongoing clinical trials can be found on clinicaltrails.gov. In addition, gene therapy is being evaluated in animal models. Periodic evaluations for cardiac, endocrine, renal and audiological complications are needed. Multidisciplinary approach (audiologist, neurologist, ophthalmologist, cardiologist, physical therapy and rehabilitation specialist) is needed for optimal management.

TABLE 11.1.4
Systemic Manifestations in LS and Their Management

Organ System	Manifestations	Management
Constitutional	Failure to thrive	Optimize nutrition, gastrostomy tube feeding
Ophthalmology	Optic atrophy, retinopathy	Visual rehabilitation
Audiology	Sensorineural deafness	Cochlear implant, hearing aids
Neurology	Dystonia, hypotonia, ataxia, seizures, peripheral neuropathy	Anticonvulsant, muscle relaxant
Musculoskeletal	Myopathy	Physical therapy, occupational therapy
Cardiovascular	Cardiomyopathy	Anticongestive therapy
Gastrointestinal	Hepatomegaly, liver disease	Avoidance of hepatotoxic medication
Respiratory	Respiratory muscle weakness	Ventilator support
Renal	Tubulopathy, chronic renal failure	Periodic monitoring

TABLE 11.1.5
Specific Therapies in LS

Condition (*gene*)	Clinical Presentation	Treatment
Coenzyme Q10 deficiency (*PDSS2*)	LS, nephropathy	Coenzyme Q10
Biotin/thiamin responsive basal ganglia disease (*SLC19A3*)	Recurrent subacute encephalopathy manifesting as confusion, ataxia, dystonia, ophthalmoplegia, and bilaterally symmetrical lesions in the brain stem and basal ganglia	Biotin and high dose thiamine
Biotinidase deficiency (*BTD*)	Psychomotor regression, hypotonia, seizures, ataxia, skin rash, alopecia, sensorineural hearing loss	Biotin
Pyruvate dehydrogenase deficiency (*PDHA1*)	Psychomotor delay and regression, hypotonia, agenesis of corpus callosum, lactate: pyruvate ratio < 20	Thiamine, ketogenic diet

TABLE 11.1.6
Research Trials in LS

Drug/Agent	Mechanism of Action	Remarks
EPI 743	It is a vitamin E analog that has strong antioxidant properties and increases intracellular reduced glutathione level.	Studies are currently underway. Initial results are encouraging.
RP 103	It is cysteamine bitartrate which reacts with cysteine in lysosome forming free cysteine and cysteine-cysteamine. Free cysteine is then transported out to the cytoplasm where it acts as a precursor of glutathione.	A long term study of RP 103 in mitochondrial diseases has been completed.

SUGGESTED READING

1. Lake NJ, Compton AG, Rahman S, Thorburn DR. Leigh syndrome: one disorder, more than 75 monogenic causes. *Ann Neurol.* 2016;79:190−203.
2. Gerards M, Sallevelt SC, Smeets HJ. Leigh syndrome: resolving the clinical and genetic heterogeneity paves the way for treatment options. *Mol Genet Metabol.* 2016;117: 300−312.
3. Thorburn DR, Rahman J, Rahman S. Mitochondrial DNA-associated Leigh syndrome and NARP. In: Adam MP, Ardinger HH, Pagon RA, et al., eds. *GeneReviews®* [*Internet*]. Seattle (WA): University of Washington, Seattle; October 30, 2003:1993−2018.
4. Rahman S, Thorburn D. Nuclear gene-encoded Leigh syndrome overview. In: Adam MP, Ardinger HH, Pagon RA, et al., eds. *GeneReviews® [Internet]*. Seattle (WA): University of Washington, Seattle; October 1, 2015:1993−2018.

Mitochondrial Disease of Nuclear Origin: Disorders of Mitochondrial DNA Replication

ABSTRACT

Mitochondrial DNA consistently replicates to maintain the mtDNA pool inside a cell. The mtDNA replication machinery is entirely nucleus derived. In addition, for replication of mtDNA, nucleotide pool needs to be maintained inside mitochondria. As mitochondria are impermeable to nucleotides, the supply of nucleotides is maintained either by recycling of nucleotides (salvage pathway) inside mitochondria or import from the cytoplasm by specific transporters on inner mitochondrial membrane. Thus, mitochondrial disease can be due to mutations in nuclear genes involved in mtDNA replication, salvage pathway, or mitochondrial nucleotide import.

KEYWORDS

Mitochondrial DNA depletion syndrome; Mitochondrial DNA maintenance; MNGIE; Multiple deletion; POLG.

BACKGROUND

There are hundreds of mitochondria in a cell, and each cell has 2–10 copies of mtDNA. Mitochondria are dynamic organelles and undergo constant fission and fusion. Damaged mitochondria are removed by this process and replaced by healthy mitochondria. Thus, unlike nuclear DNA, mtDNA consistently replicates to maintain the mtDNA pool inside a cell. The mtDNA replication machinery is entirely nucleus derived. In addition, for replication of mtDNA, nucleotide pool needs to be maintained inside mitochondria. As mitochondria are impermeable to nucleotides, the supply of nucleotide is maintained either by recycling of nucleotides (salvage pathway) inside mitochondria or import from the cytoplasm by specific transporters on inner mitochondrial membrane. Thus, mitochondrial disease can be due to mutations in nuclear genes

related to mtDNA replication, salvage pathway, or mitochondrial nucleotide import.

Defects in mtDNA replication, nucleotide salvage, or import pathways lead to the impaired synthesis of mtDNA that manifests as quantitative or qualitative defects of mtDNA. The severe forms usually present in infancy. There is a marked decline in mtDNA content (mitochondrial DNA depletion). The less severe forms tend to present later. Mitochondrial DNA shows multiple deletions in these cases. However, the severe infantile form may also show multiple deletions. Multiple deletions and depletion of mtDNA reflect a deficiency in mtDNA replication mechanism and are hallmarks of disorders of mtDNA maintenance.

Replication of mitochondrial DNA: The replication starts with separation of the double strand by helicase (*TWNK*). After this, the single strand is stabilized by mitochondrial single-strand binding protein (*SSBP1*). The polymerase complex consisting of one polymerase gamma (*POLG1*) and two accessory subunits coded by *POLG2* replicate mitochondrial DNA. In addition, mitochondrial RNA polymerase (*POLRMT*) and mitochondrial transcription factor A (*TFAM*) are needed to make RNA primer to enable the polymerase gamma to initiate replication. Mitochondrial DNA replication starts at the heavy strand origin of replication and once it reaches light strand origin of replication, replication of light strand starts in the opposite direction. Once the replication approaches near completion, RNA primer is removed by RNaseH1, DNA2, FEN1, and MGME1. DNA ligase then seals the nick, and replication is complete.

Mitochondrial nucleotide pool is maintained by conversion on deoxyribonucleosides to deoxyribonucleotides (dNTPs) via salvage pathway and import of dNTPs from cytosol to mitochondria by membrane transporters.

Mitochondrial salvage pathway: Salvage pathway consists of pyrimidine and purine salvage pathways. Thymidine kinase (*TK2*) converts pyrimidines deoxycytidine and thymidine to deoxycytidine monophosphate and thymidine monophosphate, respectively. Deoxyguanosine kinase (*DGUOK*) converts purines deoxyguanosine and deoxyadenosine to their respective monophosphates. The pyrimidine and purine monophosphates are then converted to diphosphates and triphosphates (dNTPs) by sequential actions of nucleotide monophosphate kinase (*NMPK*) and nucleotide diphosphate kinase (*NDPK*), respectively.

Mitochondrial nucleotide import: Adenine nucleotide translocator (*ANT*) present in mitochondrial inner membrane translocates ATP to cytoplasm from the mitochondrial matrix in exchange of ADP. ADP can then be converted to dADP and dATP in the mitochondria. ANT is anchored to inner mitochondrial membrane phospholipids particularly cardiolipin. Thus, maintaining the integrity of the inner mitochondrial membrane is essential for ANT function.

Cytoplasmic salvage pathway: Maintenance of adequate cytoplasmic nucleotide pool is vital to supply nucleotides to the mitochondria. The enzyme ribonucleotide reductase (*RNR*) catalyzes the conversion of ribonucleotide diphosphates to deoxyribonucleotide diphosphates that are substrates for DNA replication. RNR is a tetramer made of 2, R1 and 2, R2 (in dividing cells) or p53R2 (in postmitotic cells). Mitochondrial DNA replication occurs in both dividing and postmitotic cells. Hence, RNR (R1/p53R2) plays a direct role in maintaining nucleotide pool in the cytoplasm and indirectly maintains mitochondrial nucleotide pool in postmitotic cells. *RRM2B* encodes p53R2.

CLINICAL PRESENTATION

Disorders of mitochondrial DNA maintenance present with a wide spectrum of clinical manifestations. They may first present in infancy or adult life. Salient features of the disorders of mitochondrial DNA maintenance due to defects in replication machinery and maintenance of nucleotide pool are summarized in Tables 12.1 and 12.2.

GENETICS

Mitochondrial DNA maintenance defects that are caused by nuclear gene mutations are inherited in autosomal fashion. Autosomal recessive is the most common mode of inheritance. However, autosomal dominant inheritance is also found (Table 12.3). Occasionally, the autosomal dominant condition is sporadic due to de novo mutation.

DIAGNOSIS

Diagnosis can be made by targeted genetic testing if the clinical picture is characteristic such as in MNGIE or *MPV17*-related Navajo hepatopathy. Skeletal muscle biopsy provides very useful information. In mtDNA, maintenance defects, apart from combined OXPHOS deficiency, multiple deletions, and/or depletion of mtDNA, are seen. However, the absence of these findings does not rule out a mitochondrial DNA maintenance disorder. Different molecular genetic diagnostic approaches are summarized in Table 12.4.

Blood lactate and cerebrospinal fluid lactate are often elevated. Magnetic resonance imaging (MRI) of the brain may show lesions in white matter, basal ganglia, or brain stem. Magnetic resonance spectroscopy of brain parenchymal tissue may show lactate peak.

MANAGEMENT

As mitochondrial DNA maintenance disorders are multisystemic diseases, diagnosis of mitochondrial DNA maintenance disorder should lead to an evaluation for other systemic manifestations. Management is mainly symptomatic, avoidance of mitotoxic medications, physical and occupational therapy (Table 12.5). Mitochondrial supplements such as coenzyme Q10, L-carnitine, alpha lipoic acid, and riboflavin can be tried. Bone marrow transplantation can be considered for *TYMP*-related MNGIE, while liver transplantation can be an option for *DGUOK*-related hepatocerebral mitochondrial DNA depletion syndrome. For research trails, clinicaltrials.gov is a very useful resource. Periodic evaluations for cardiac, neurological, ophthalmologic, and audiological complications are needed. A multidisciplinary approach (audiologist, neurologist, ophthalmologist, endocrinologist, cardiologist, gastroenterologist, physical therapy, and rehabilitation specialist) is needed on an individual case basis for optimal management.

TABLE 12.1

Genes Related to Mitochondrial DNA Replication and Associated Clinical Phenotypes

Gene	Pathophysiology	Clinical Syndrome	mtDNA Defect	Clinical Manifestations
POLG	Polymerization of mitochondrial DNA	Childhood myocerebrohepatopathy spectrum (MCHS)	Depletion	Neonatal/infantile presentation Hypotonia Developmental delay Liver disease
		Alpers syndrome	Depletion	Early childhood presentation Intractable epilepsy Psychomotor regression Liver disease
		Ataxia neuropathy spectrum (ANS)	Multiple deletions	Adolescence/young adulthood presentation Cerebellar ataxia Sensory ataxia Peripheral axonal neuropathy
		Myoclonic epilepsy myopathy sensory ataxia (MEMSA)	Multiple deletions	Adolescence/young adulthood presentation Myoclonus Epilepsy Myopathy Sensory ataxia
		Autosomal recessive PEO	Multiple deletions	Adolescence/young adulthood presentation Ptosis Ophthalmoplegia Mood disorder Parkinsonism Ataxia
		Autosomal dominant PEO	Multiple deletions	Adulthood presentation Ptosis Ophthalmoplegia Myopathy Parkinsonism Mood disorder
		Mitochondrial neurogastrointestinal encephalopathy (MNGIE)-like phenotype	Depletion and multiple deletions	Gastrointestinal dysmotility Myopathy Neuropathy
POLG2	POLG2 is an accessory subunit of the polymerase gamma complex which enhances the polymerase and exonuclease functions and increases processivity of the polymerase enzyme	Autosomal dominant PEO	Multiple deletions	Infancy to adulthood presentation Ptosis Ophthalmoplegia Proximal myopathy Exercise intolerance

Continued

TABLE 12.1
Genes Related to Mitochondrial DNA Replication and Associated Clinical Phenotypes—cont'd

Gene	Pathophysiology	Clinical Syndrome	mtDNA Defect	Clinical Manifestations
TWNK (PEO1)	Unwinds the DNA double helix needed for replication	Infantile onset spinocerebellar ataxia (IOSCA)	Depletion	Infancy/early childhood presentation Ataxia Hypotonia Hyporeflexia Ophthalmoplegia Sensorineural hearing loss Progression to epileptic encephalopathy in adolescence Seen in the Finnish population
		Hepatocerebral syndrome	Depletion	Neonatal/infantile presentation Liver disease Lactic acidosis Hypotonia Developmental delay Seizures
		Autosomal dominant PEO	Multiple deletions	Adulthood presentation Ptosis Ophthalmoplegia Proximal myopathy Exercise intolerance
TFAM	Generates RNA primer required for initiation of mtDNA replication	Neonatal liver failure	Depletion	Neonatal presentation Hypoglycemia Liver failure
RNASEH1	Degrades the RNA in RNA/DNA hybrid at the replication initiation site, thus allowing complete mtDNA replication.	Autosomal recessive PEO	Multiple deletions and Depletion	Adulthood presentation Ptosis Ophthalmoplegia Muscle weakness Dyspnea
DNA2	Acts in concert with FEN1 to remove the flap intermediate formed as a result of DNA replication reaching the initiation site. This enables complete mtDNA replication	Autosomal dominant PEO	Multiple deletions	Adulthood presentation Ptosis Ophthalmoplegia Myopathy Exercise intolerance
MGME1	Cleaves the single-stranded DNA in flap intermediate formed as a result of DNA replication reaching the initiation site. This enables complete mtDNA replication	Autosomal recessive PEO	Multiple deletions and Depletion	Childhood/early adulthood presentation Ptosis Ophthalmoplegia Generalized muscle wasting Respiratory failure

	TABLE 12.2			
	Nuclear Genes Related to Maintenance of Mitochondrial Nucleotide Pool and Associated Clinical Phenotypes			
Gene	**Pathophysiology**	**Clinical Syndrome**	**mtDNA Defect**	**Clinical Manifestations**
TK2	TK2 catalyzes the conversion of pyrimidines, deoxycytidine, and thymidine to deoxycytidine monophosphate, and thymidine monophosphate, respectively.	Autosomal recessive myopathic mitochondrial DNA depletion syndrome	Depletion	Infantile/childhood presentation: Progressive muscle weakness Respiratory failure Bulbar weakness
		Autosomal recessive PEO	Multiple deletions	Adulthood presentation Ptosis Ophthalmoplegia Proximal myopathy
DGUOK	Deoxyguanosine kinase converts purines, deoxyguanosine, and deoxyadenosine, to their respective monophosphates.	Hepatocerebral syndrome	Depletion	Neonatal/Infantile presentation: Progressive liver disease, hypotonia, developmental delay Childhood presentation: Progressive liver disease
		Autosomal recessive PEO	Multiple deletions	Adulthood presentation Ptosis Ophthalmoplegia Proximal myopathy
SUCLA2 SUCLG1	SUCLA2 and SUCLG1 are subunits of succinyl coA ligase (SUCL) and forms complex with mitochondrial nucleotide diphosphate kinase (NDPK), which converts nucleotide diphosphates to triphosphates. SUCL stabilizes the NDPK complex	Encephalomyopathy Leigh/Leigh-like syndrome	Depletion	Neonatal to early childhood presentation Hypotonia Muscle atrophy Dystonia Psychomotor delay Basal ganglia hyperintensities Elevated methylmalonic acid (due to elevated succinyl coA) SUCLG1: earlier onset and more severe presentation, Hypertrophic cardiomyopathy and liver failure in addition to above
ANT1 (SLC25A4)	Adenine nucleotide translocator (ANT) translocates ATP to cytoplasm from the mitochondrial matrix in exchange of ADP. ADP can then be converted to dADP and dATP in the mitochondria	Cardiomyopathic mitochondrial DNA depletion syndrome	Multiple deletions	Early childhood presentation Hypertrophic cardiomyopathy Myopathy Lactic acidosis
		Autosomal dominant PEO	Multiple deletions	Adulthood presentation Ptosis Ophthalmoplegia Myopathy Exercise intolerance

Continued

TABLE 12.2
Nuclear Genes Related to Maintenance of Mitochondrial Nucleotide Pool and Associated Clinical Phenotypes—cont'd

Gene	Pathophysiology	Clinical Syndrome	mtDNA Defect	Clinical Manifestations
MPV17	Unknown function, presumably involved in the import of nucleotides in mitochondria	Hepatocerebral mitochondrial DNA depletion syndrome	Depletion	Neonatal to early childhood presentation Progressive liver disease Sensorimotor neuropathy Hypotonia White matter hyperintensities Common in the Navajo population
		Neuromyopathy	Multiple deletions	Adolescence to adulthood presentation Myopathy Sensory neuropathy
TYMP	Thymidine phosphorylase (TP) converts thymidine and deoxyuridine to thymine and uridine, respectively. TP deficiency causes an increase of thymidine and deoxyuridine and nucleotide pool imbalance leading to mtDNA instability, deletion, and depletion	Mitochondrial neurogastrointestinal encephalopathy (MNGIE)	Point mutations Multiple deletions and Depletion	Adolescence to adulthood presentation Gastrointestinal dysmotility (diarrhea, dysphagia, abdominal distention) Cachexia Ptosis Ophthalmoplegia Sensorimotor neuropathy Leukoencephalopathy Myopathy
RRM2B	RRM2B encodes p53R2 which is a subunit of RNR complex. RNR converts ribonucleotides to deoxyribonucleotides required for mtDNA replication	Encephalomyopathic mitochondrial DNA depletion syndrome with renal tubulopathy	Depletion	Infantile presentation Developmental delay Hypotonia Proximal renal tubulopathy Nephrocalcinosis Lactic acidosis Failure to thrive
		MNGIE-type mitochondrial depletion syndrome	Depletion	Adulthood presentation Gastrointestinal dysmotility (diarrhea, dysphagia, abdominal distention) Ptosis Ophthalmoplegia Myopathy White matter hyperintensities
		Autosomal recessive PEO	Multiple deletions	Childhood presentation Ptosis Ophthalmoplegia Sensorineural hearing loss
		Autosomal dominant PEO	Multiple deletions	Adulthood presentation Ptosis Ophthalmoplegia Sensorineural hearing loss

TABLE 12.3
Inheritance Pattern of Mitochondrial DNA Maintenance Disorders

Inheritance Pattern	Gene	Clinical Phenotype
Autosomal dominant	*POLG* *POLG2* *ANT1* *PEO1* *RRM2B* *DNA2*	Autosomal dominant PEO
Autosomal recessive	All other genes	Tables 1 and 2

TABLE 12.4
Molecular Genetic Diagnostic Approaches

Molecular Genetic Test	Remarks
Targeted nuclear gene sequencing	*TYMP*: gastrointestinal dysmotility, cachexia, PEO, white matter hyperintensities *MPV17:* Progressive liver disease and neuropathy in an infant/child of the Navajo population
Targeted gene panel	If there is evidence of mitochondrial DNA maintenance defect in muscle biopsy such as multiple deletion/depletion, a panel of genes associated with mitochondrial DNA maintenance defect can be assayed first.
Combined mitochondrial genome and nuclear gene panel	A popular approach is to combine mtDNA sequencing with the sequencing of known nuclear genes related with mitochondrial function. If family history is suggestive of autosomal inheritance, only nuclear gene panel can be ordered first.
Whole exome/genome sequencing with mitochondrial genome sequencing	This is the most comprehensive approach. It not only allows diagnosis of a mitochondrial disorder but also other disorders which are in the differential diagnosis. However, there is also an increased likelihood of finding variants of unknown significance.

TABLE 12.5
Systemic Manifestations of Mitochondrial DNA Maintenance Disorders and Their Management

Organ System	Manifestations	Management
Constitutional	Failure to thrive	Optimize nutrition, swallowing evaluation, gastrostomy tube feeding,
Ophthalmology	Ptosis, retinopathy, optic neuropathy	Ptosis surgery, visual rehabilitation, correction glasses
Audiology	Sensorineural deafness	Cochlear implant, hearing aids, sign language
Neurology	Intellectual disability, developmental delays, hypotonia, dystonia, seizures	Anticonvulsant
Musculoskeletal	Myopathy	Physical therapy, occupational therapy
Endocrinology	Diabetes mellitus, hypothyroidism	Diet, insulin, thyroid replacement etc.
Cardiovascular	Cardiomyopathy	Angiotensin-converting enzyme inhibitors
Respiratory	Respiratory muscle weakness	Ventilator support
Gastrointestinal	Dysphagia, gastroparesis, liver disease	Motility agents, avoidance of hepatotoxic agents

SUGGESTED READING

1. Stumpf JD, Saneto RP, Copeland WC. Clinical and molecular features of POLG-related mitochondrial disease. *Cold Spring Harb Perspect Biol.* 2013;5:a011395.
2. Nogueira C, Almeida LS, Nesti C, et al. Syndromes associated with mitochondrial DNA depletion. *Ital J Pediatr.* 2014;40:34.
3. El-Hattab AW, Craigen WJ, Scaglia F. Mitochondrial DNA maintenance defects. *Biochim Biophys Acta.* 2017;1863: 1539−1555.
4. Rusecka J, Kaliszewska M, Bartnik E, Tońska K. Nuclear genes involved in mitochondrial diseases caused by instability of mitochondrial DNA. *J Appl Genet.* 2018;59:43−57.

CHAPTER 13

Mitochondrial Disease of Nuclear Origin: Disorders of Mitochondrial DNA Transcription and Translation

ABSTRACT

Mitochondrial DNA transcription is polycistronic. It originates at the heavy and light strand promoters and progresses around almost the entire length of DNA strand. Individual m-RNA, t-RNA, and r-RNAs are then cleaved from this polycistronic transcript, and they undergo further modification before participating in translation. Mitochondrial disease can be a result of a defect in any of these steps.

KEYWORDS

Aminoacylation; Mitochondrial disease; Polycistronic; Transcription; Translation.

BACKGROUND

Mitochondrial DNA transcription is polycistronic. It originates at the heavy and light strand promoters and progresses around almost the entire length of the DNA strand. Individual m-RNA, t-RNA, and r-RNAs are then cleaved from this polycistronic transcript, and they undergo further modification before participating in translation. Mitochondrial disease can be a result of a defect in any of these steps.

CLINICAL PRESENTATION

Mitochondrial DNA Transcription and Associated Disorders

Each strand of the mitochondrial DNA is transcribed as a pre-RNA consisting of protein-coding regions (m-RNA) and ribosomal RNAs flanked by t-RNAs. Transcription is initiated at the promoter regions of heavy and light strands by the mitochondrial RNA polymerase (*POLRMT*) in the presence of the transcription factors, TFAM and TFB2M. A mature transcript is terminated by mitochondrial termination factor 1 (*MTERF1*). TFAM (mitochondrial transcription factor A) is essential for mtDNA replication, transcription, and nucleoid

packaging. Disorder of mitochondrial transcription is summarized in Table 13.1.

Mitochondrial Pre-RNA Processing and Associated Disorders

The pre-RNA transcript consists of m-RNAs and r-RNAs flanked by t-RNAs. The t-RNAs are then cleaved by RNase P and RNase Z at the $5'$- and $3'$-ends, respectively. RNase P complex is formed by mitochondrial RNase P protein 1 (MRPP1), MRPP2, and MRPP3. RNase Z is encoded by *ELAC2*. A defect in mitochondrial pre-RNA processing leads to decreased t-RNA levels, increased pre-RNA, and generalized mitochondrial dysfunction. Table 13.2 summarizes disorders of mitochondrial pre-RNA processing.

Mitochondrial t-RNA Modification and Associated Disorders

After being spliced out from the precursor RNA, t-RNAs undergo modification. Maturation of t-RNAs requires the addition of CCA sequence to the $3'$-terminal end by t-RNA nucleotidyltransferase (*TRNT1*). There are 22 t-RNAs in mitochondria to decode 60 genetic codes in mitochondrial m-RNA. Although a genetic code is specific for an amino acid, an amino acid is coded by more than one genetic code. This is possible because

TABLE 13.1

Nuclear Gene Related to Mitochondrial DNA Transcription and Associated Clinical Phenotype

Gene	Pathophysiology	Clinical Manifestations
TFAM	mtDNA transcription, translation, and nucleoid packaging	Neonatal/Infantile presentation Intrauterine growth retardation Progressive fatal liver failure

Mitochondrial Medicine. https://doi.org/10.1016/B978-0-12-817006-9.00013-7

TABLE 13.2
Nuclear Genes Related to Mitochondrial pre-RNA Processing and Associated Clinical Phenotypes

Gene	Pathophysiology	Clinical Manifestations
MRPP1 (TRMT10C)	Forms part of RNase P complex that cleaves t-RNAs from the precursor RNA at the 5′-end	Infantile presentation Hypotonia Feeding difficulties Sensorineural hearing loss Lactic acidosis
MRPP2 (HSD17B10/HADH2)	Forms part of RNase P complex that cleaves t-RNAs from the precursor RNA at the 5′-end MRPP2 is a multifunctional protein and is involved in isoleucine, short chain hydroxy-coA, and 17 β-hydroxysteroid metabolism	Infantile to early childhood presentation Psychomotor delay and regression Seizures Choreoathetosis Spasticity Retinal degeneration Sensorineural hearing loss Hypertrophic cardiomyopathy 2-Methyl, 3-hydroxybutyric aciduria Also known as 2-methyl, 3-hydroxybutyryl coA dehydrogenase deficiency (MHBD)
ELAC2	Cleaves t-RNAs from the precursor RNA at the 3′-end	Infantile presentation Severe hypertrophic cardiomyopathy Hypotonia Failure to thrive Lactic acidosis

the base pairing between the first position of t-RNA anticodon (wobble position) and the last position of the codon is less precise allowing flexibility. Although modification at the wobble position (position 34) allows flexibility, modifications at the 3′-adjacent position in the anticodon loop (position 37) ensure transcriptional fidelity. Thus frameshifting errors are avoided. In addition, t-RNA modifications are needed to stabilize its secondary structure. Table 13.3 summarizes disorders of mitochondrial t-RNA modification.

Each t-RNA is charged with its corresponding amino acid by aminoacyl t-RNA synthetase. There are 19 mitochondrial aminoacyl t-RNA synthetases. Glycyl and lysl t-RNA synthetases aminoacylate both cytosolic and mitochondrial t-RNAs. There is no mitochondrial glutaminyl t-RNA synthetase. Glutamine t-RNA is formed indirectly in two steps. In the first step, glutamate is added by glutamyl t-RNA synthetase that then undergoes transamidation to form glutamine t-RNA. The methionyl t-RNA (t-RNA^met) destined for translation initiation undergoes formylation by mitochondrial methionyl t-RNA formyltransferase (MTFMT) to produce formyl methionine t-RNA^met essential for translation initiation. Table 13.4 summarizes disorders of mitochondrial t-RNA aminoacylation.

Mitochondrial m-RNA Processing and Associated Disorders

Mitochondrial m-RNAs undergo polyadenylation. Polyadenylation completes the UAA stop codon on most m-RNAs and also maintains the integrity of m-RNA by preventing degradation. Leucine-rich PPR motif-containing protein (LRPPRC) stabilizes the secondary structure of m-RNA to enable polyadenylation and coordinated translation. Table 13.5 summarizes disorders of mitochondrial m-RNA processing.

Mitochondrial Ribosomal Structure, Assembly, and Associated Disorders

Translation of mitochondrial m-RNA occurs at the mitochondrial ribosome. Mitoribosome is a 55S unit composed of two subunits: small 28S and large 39S. The 28S subunit is made by 12S r-RNA and 30 ribosomal proteins. The 39S subunit is made of 16S r-RNA, 52 ribosomal protein, and mitochondrial t-RNA for valine or phenylalanine. The r-RNAs provide the scaffold for ribosomal assembly and undergo several modifications. Disorders of mitoribosome structure and assembly are summarized in Table 13.6.

TABLE 13.3
Nuclear Genes Related to Mitochondrial t-RNA Modification and Associated Clinical Phenotypes

Gene	Pathophysiology	Clinical Manifestations
TRNT1	Maturation of the 3′-end of t-RNA by addition of CCA	Neonatal to juvenile presentation Sideroblastic anemia B-cell immunodeficiency Periodic fever Developmental delays Erythrocytic microcytosis Retinitis pigmentosa
TRMU	Thiolation of the uridine nucleoside at the wobble position of the t-RNA for glutamate, glutamine, and lysine. This requires cysteine	Infantile presentation Reversible infantile liver failure. Cysteine is an essential amino acid in infants because of the low cystathionase activity in infants
MTO1	Taurinomethylation of the uridine nucleoside at the wobble position of the t-RNA for glutamate, glutamine, and lysine	Infantile presentation Hypertrophic cardiomyopathy Developmental delay Hypotonia Lactic acidosis
GTPBP3	GTPBP3-MTO1 jointly catalyzes taurinomethylation of the uridine nucleoside at the wobble position of the t-RNA for glutamate, glutamine, and lysine	Early childhood presentation Hypertrophic cardiomyopathy Developmental delay Hypotonia Lactic acidosis Bilateral lesions in the brain stem and basal ganglia
NSUN3	NSUN3 modifies the cytosine at the wobble position of t-RNA for methionine enabling it to bind 3 codons for methionine.	Infantile presentation Developmental delay Microcephaly Failure to thrive Hypotonia External ophthalmoplegia Nystagmus Lactic acidosis
TRMT5	t-RNA methyltransferase 5 methylates guanosine at position 37 of many t-RNAs to prevent frameshifting errors	Childhood to adult presentation Failure to thrive Hypertrophic cardiomyopathy Exercise intolerance Lactic acidosis
TRIT1	t-RNA isopentenyltransferase 1 adds isopentenyl group to the adenosine at position 37 of many cytoplasmic and mitochondrial t-RNAs to prevent frameshifting errors	Childhood presentation Microcephaly Developmental delay Seizures
PUS1	Pseudouridylate synthase 1 catalyzes isomerization of uridine to pseudouridine at several positions in cytoplasmic and mitochondrial t-RNAs required for stability of the secondary structure	Childhood to adult presentation Myopathy, lactic acidosis, and sideroblastic anemia (MLASA)

TABLE 13.4

Nuclear Genes Related to Mitochondrial t-RNA Aminoacylation and Associated Clinical Phenotypes

Gene	Pathophysiology	Clinical Manifestations
AARS2	AARS2 conjugates alanine with the mitochondrial t-RNA for alanine	Infantile and adulthood presentation Infantile—failure to thrive, hypertrophic cardiomyopathy, lactic acidosis Adulthood—tremor, ataxia, leukoencephalopathy, ovarian failure in women
CARS2	CARS2 conjugates cysteine with the mitochondrial t-RNA for cysteine	Infantile to early childhood presentation Developmental delay/regression Epileptic encephalopathy Visual failure Hearing loss Cerebral atrophy White matter changes on brain MRI
DARS2	DARS2 conjugates aspartate with the mitochondrial t-RNA for aspartic acid	Early childhood to juvenile presentation LSBL (leukoencephalopathy with **s**pinal cord and **b**rain stem involvement and **l**actate elevation) Ataxia Tremor Spasticity Nystagmus Delayed development Peripheral neuropathy White matter lesions in the periventricular, brain stem, cerebellum, and spinal cord tracts.
EARS2	EARS2 conjugates glutamate with the mitochondrial t-RNA for glutamate	Infantile presentation LTBL (leukoencephalopathy with **t**halamus and **b**rain stem involvement and **l**actate elevation) Biphasic presentation. Early infantile—neurological regression, visual impairment, seizures, spasticity followed by clinical stagnation Late infantile presentation—neurological regression, irritability followed by improvement.
FARS2	FARS2 conjugates phenylalanine with the mitochondrial t-RNA for phenylalanine	Infantile presentation Developmental delay Microcephaly Hearing impairment Seizures Cerebral atrophy Lactic acidosis
HARS2	HARS2 conjugates histidine with the mitochondrial t-RNA for histidine	Childhood presentation Perrault syndrome (hearing loss, ovarian failure in female)
IARS2	IARS2 conjugates isoleucine with the mitochondrial t-RNA for isoleucine	Infantile and adulthood presentation Leigh syndrome CAGSSS syndrome (**ca**taract, **g**rowth hormone deficiency, **s**ensory neuropathy, **s**ensorineural deafness **s**keletal dysplasia **s**yndrome)
LARS2	LARS2 conjugates leucine with the mitochondrial t-RNA for leucine	Congenital presentation: HLASA (**h**ydrops, **l**actic **a**cidosis, and **s**ideroblastic **a**nemia) Juvenile presentation: Perrault Syndrome (hearing loss, ovarian failure in female)

(continued)

TABLE 13.4

Nuclear Genes Related to Mitochondrial t-RNA Aminoacylation and Associated Clinical Phenotypes—cont'd

Gene	Pathophysiology	Clinical Manifestations
MARS2	MARS2 conjugates methionine with the mitochondrial t-RNA for methionine	Childhood to adulthood onset ARSAL (**a**utosomal **r**ecessive **s**pastic **a**taxia with **l**eukoencephalopathy)
NARS2	NARS2 conjugates asparagine with the mitochondrial t-RNA for asparagine	Infantile presentation Neurodegeneration Leigh syndrome Liver involvement Sensorineural hearing loss
PARS2	PARS2 conjugates proline with the mitochondrial t-RNA for proline	Infantile presentation Neurodegeneration Leigh syndrome Liver involvement Sensorineural hearing loss
RARS2	RARS2 conjugates arginine with the mitochondrial t-RNA for arginine	Neonatal presentation Pontocerebellar hypoplasia Failure to thrive Microcephaly Hypotonia Seizures
SARS2	SARS2 conjugates serine with the mitochondrial t-RNA for serine	Infantile presentation HUPRA (**h**yper**u**ricemia, **p**ulmonary hypertension, **r**enal failure, **a**lkalosis) syndrome Global developmental delays
TARS2	TARS2 conjugates threonine with the mitochondrial t-RNA for threonine	Neonatal presentation Hypotonia Thin corpus callosum Hyperintense lesions in globus pallidus
VARS2	VARS2 conjugates valine with the mitochondrial t-RNA for valine	Neonatal presentation Severe encephalomyopathy Cardiomyopathy Abnormal brain MRI Lactic acidosis
WARS2	WARS2 conjugates tryptophan with the mitochondrial t-RNA for tryptophan	Infantile or early childhood onset Infantile-onset encephalopathy Infantile-onset Parkinsonism Intellectual disability
YARS2	YARS2 conjugates tyrosine with the mitochondrial t-RNA for tyrosine	Infantile or early childhood onset MLASA (myopathy, lactic acidosis, sideroblastic anemia) type 2
GARS	GARS2 conjugates glycine with the cytoplasmic and mitochondrial t-RNA for glycine	Adult-onset (autosomal dominant): Charcot-Marie-Tooth disease type 2D, distal hereditary motor neuropathy type VA. Childhood onset (autosomal recessive): exercise intolerance, cardiomyopathy, lactic acidosis, white matter changes in brain MRI.
KARS	KARS2 conjugates lysine with the cytoplasmic and mitochondrial t-RNA for lysine	Infantile to adulthood onset Autosomal recessive deafness Charcot-Marie-Tooth disease, intermediate recessive B
MTFMT	Formylation of methionine attached to t-RNA^{met} destined for translation initiation that increases its efficiency toward mitochondrial initiation factor 2	Childhood presentation Developmental delay Learning disabilities Lesions on brain imaging resembling Leigh syndrome

TABLE 13.5
Nuclear Genes Related to Mitochondrial m-RNA Processing and Associated Clinical Phenotypes

Gene	Pathophysiology	Clinical Manifestations
MTPAP	Polyadenylation of mitochondrial mRNA	Early childhood to juvenile presentation Progressive spastic ataxia Optic atrophy More common in Old Order Amish population
LRPPRC	Acts as mitochondrial m-RNA chaperone and stabilizes the secondary structure of m-RNA	Infantile presentation Leigh syndrome French Canadian type Psychomotor regression Intermittent lactic acidosis and coma Bilateral basal ganglia and brain stem lesions.
PNPT1	Processing of m-RNA transcripts once it is cleaved from the polycistronic transcript	Infantile presentation Leigh syndrome

TABLE 13.6
Nuclear Genes Related to Mitochondrial Ribosome Structure and Assembly and Associated Clinical Phenotypes

Gene	Pathophysiology	Clinical Manifestations
MRPL3	Encodes a mitoribosome large subunit (39S) ribosomal protein	Infantile presentation Developmental delay Hypertrophic cardiomyopathy
MRPL12	Encodes a mitoribosome large subunit (39S) ribosomal protein	Infantile presentation Psychomotor regression Failure to thrive
MRPL44	Encodes a mitoribosome large subunit (39S) ribosomal protein	Infantile presentation Hypertrophic cardiomyopathy
MRPS16	Encodes a mitoribosome small subunit (28S) ribosomal protein	Neonatal presentation Small for gestational age Dysmorphic facies Hypotonia Agenesis of corpus callosum Lactic acidosis
MRPS22	Encodes a mitoribosome small subunit (28S) ribosomal protein	Neonatal presentation Cornelia de Lange-like phenotype Hypotonia Seizures Hypoplasia of corpus callosum Hypertrophic cardiomyopathy
MRPS34	Encodes a mitoribosome small subunit (28S) ribosomal protein	Neonatal presentation Leigh syndrome
MRM2	MRM2 causes 2′-O-methyl modification of 16S r-RNA	Childhood presentation Progressive encephalomyopathy Stroke-like episodes
ERAL1	Required for proper assembly of 28S subunit	Childhood to adult presentation Perrault syndrome (hearing loss, ovarian failure in female)

TABLE 13.7
Nuclear Genes Related to Mitochondrial Translation Initiation, Elongation, and Termination and Associated Clinical Phenotypes

Gene	Pathophysiology	Clinical Manifestations
GFM1	Encodes mitochondrial translation elongation factor G1	Neonatal presentation Psychomotor delay Hypotonia Seizures Hypoplasia of corpus callosum Liver failure
TUFM	Encodes mitochondrial translation elongation factor Tu	Neonatal presentation Psychomotor delay Leigh syndrome
TSFM	Encodes mitochondrial translation elongation factor Ts	Infantile-onset cardiomyopathy Juvenile-onset Leigh syndrome
C120RF65	Encodes a protein that is critical for release of newly synthesized protein from the translation site	Infantile tor early childhood presentation: psychomotor delay and regression, optic atrophy, nystagmus, lesions in basal ganglia and brain stem Childhood presentation: spastic paraplegia with optic atrophy and axonal neuropathy (SPG55)
RMND1	RMND1 is an integral inner mitochondrial membrane protein that stabilizes mitochondrial ribosome for translation to occur	Neonatal or infantile presentation Hypotonia Seizures Myopathy Renal disease Sensorineural deafness Cardiomyopathy Lactic acidosis

Mitochondrial Translation Initiation, Elongation, Termination, and Associated Disorders

Mitochondrial translation starts with the formation of initiation complex that consists of a 28S ribosomal subunit, m-RNA, formyl methionine t-RNA, and mitochondrial initiation factors 2/3. Translation proceeds with the help of elongation factors G1, Tu, and Ts. The stop codon is recognized by the release factor releasing protein from the translation site. Table 13.7 summarizes disorders of mitochondrial translation, initiation, elongation, and termination.

GENETICS

Mitochondrial DNA transcription and translation defects caused by nuclear gene mutations are inherited in autosomal fashion. Autosomal recessive is the most common mode of inheritance. However, some disorders are inherited in autosomal dominant and X-linked manners (Table 13.8).

DIAGNOSIS

Diagnosis can be made by targeted genetic testing if the clinical picture and biochemical profile are characteristics such as in MHBD. Different molecular genetic diagnostic approaches are summarized in Table 13.9.

TABLE 13.8
Inheritance Pattern of Mitochondrial DNA Maintenance Disorders

Inheritance Pattern	Gene	Clinical Phenotype
X-linked dominant	MRPP2	2-methyl, 3-hydroxybutyryl coA dehydrogenase deficiency (MHBD)
Autosomal dominant	GARS	Charcot-Marie-Tooth disease type 2D
Autosomal recessive	All other genes	Tables 1−7

TABLE 13.9
Molecular Genetic Diagnostic Approaches

Molecular Genetic Test	Remarks
Targeted nuclear gene sequencing	*MRPP2:* Psychomotor delay and regression, seizures, choreoathetosis, spasticity, retinal degeneration, sensorineural hearing loss, hypertrophic cardiomyopathy, 2-methyl, 3-hydroxybutyric aciduria
Nuclear gene panel	If the clinical picture is consistent with a mitochondrial and family history is suggestive of autosomal inheritance, a panel of nuclear genes related to mitochondrial diseases can be ordered first.
Combined mitochondrial genome and nuclear gene panel	A popular approach is to combine mtDNA sequencing with the sequencing of known nuclear genes related with mitochondrial function.
Whole exome/genome sequencing with mitochondrial genome sequencing	This is the most comprehensive approach. It not only allows diagnosis of a mitochondrial disorder but also other disorders that are in the differential diagnosis. However, there is also an increased likelihood of finding variants of unknown significance.

TABLE 13.10
Systemic Manifestations of Mitochondrial DNA Transcription and Translation Disorders and Their Management

Organ System	Manifestations	Management
Constitutional	Failure to thrive	Optimize nutrition, swallowing evaluation, gastrostomy tube feeding
Ophthalmology	Retinopathy, optic neuropathy	Visual rehabilitation, correction glasses
Audiology	Sensorineural deafness	Cochlear implant, hearing aids, sign language
Neurology	Intellectual disability, developmental delays, hypotonia, dystonia, seizures	Anticonvulsant, muscle relaxant
Musculoskeletal	Myopathy	Physical therapy, occupational therapy
Endocrinology	Ovarian failure	Hormone replacement
Cardiovascular	Cardiomyopathy	ACE inhibitors
Respiratory	Respiratory muscle weakness	Ventilator support
Gastrointestinal	Dysphagia, gastroparesis, liver disease	Motility agents, avoidance of hepatotoxic agents
Renal	Tubulopathy, renal failure	Renal transplantation

Blood lactate and cerebrospinal fluid lactate are often elevated. Magnetic resonance imaging (MRI) of the brain may show lesions in white matter, basal ganglia, or brain stem. Magnetic resonance spectroscopy of the brain parenchymal tissue may show lactate peak.

MANAGEMENT

Mitochondrial DNA transcription and translation disorders are often multisystemic. Diagnosis of mitochondrial transcription or translation related disorder should lead to an evaluation for other systemic manifestations. Management is mainly symptomatic, avoidance of mitotoxic medications, physical and occupational therapy (Table 13.10). Mitochondrial supplements such as coenzyme Q10, L-carnitine, alpha lipoic acid, and riboflavin can be tried. For research trails, clinicaltrials.gov is a very useful resource. Periodic evaluations for cardiac, neurological, ophthalmologic, and audiological complications are needed and can be tailored on an individual basis. A multidisciplinary approach (audiologist, neurologist, ophthalmologist, cardiologist, endocrinologist, nephrologist, gastroenterologist, physical therapy, and rehabilitation specialist) is needed on an individual case basis for optimal management.

SUGGESTED READING

1. Boczonadi V, Horvath R. Mitochondria: impaired mitochondrial translation in human disease. *Int J Biochem Cell Biol.* 2014;48:77–84.

2. Powell CA, Nicholls TJ, Minczuk M. Nuclear-encoded factors involved in post-transcriptional processing and modification of mitochondrial tRNAs in human disease. *Front Genet.* 2015;6:79.

3. Boczonadi V, Ricci G, Horvath R. Mitochondrial DNA transcription and translation: clinical syndromes. *Essays Biochem.* 2018;62:321–340.

4. Bohnsack MT, Sloan KE. The mitochondrial epitranscriptome: the roles of RNA modifications in mitochondrial translation and human disease. *Cell Mol Life Sci.* 2018;75:241–260.

Mitochondrial Disease of Nuclear Origin: Disorders of Mitochondrial Homeostasis

ABSTRACT

Mitochondria are vital cellular organelles. Maintenance of mitochondrial integrity is necessary for its optimal function. To maintain a healthy mitochondrial network in a cell, mechanisms are in place to ensure proper mitochondrial biogenesis and removal of damaged mitochondria. Defects in mitochondrial homeostasis mechanisms can lead to mitochondrial dysfunction and disease. This chapter outlines the disorders of mitochondrial homeostasis.

KEYWORDS

Cardiolipin; Homeostasis; Mitochondrial dynamics; Mitochondrial fission; Mitochondrial fusion; Mitochondrial protein quality control; Proteostasis.

BACKGROUND

Maintenance of a healthy mitochondrial network is essential for its optimal function. Mitochondrial biogenesis includes transcription and translation of nuclear and mitochondrial genes. Proteins encoded by nuclear genes are then imported in mitochondria and assembled into protein complexes. Respiratory chain complex units are assembled as supercomplexes in the mitochondrial inner membrane. Biogenesis of mitochondrial membranes requires synthesis and import of membrane proteins and phospholipids. A defect in protein import or assembly can lead to impaired biogenesis and mitochondrial damage. Moreover, mitochondrial proteins are at increased risk of damage from the ROS produced by the electron transport chain. Protein quality control measures ensure removal of abnormal and damaged proteins to maintain mitochondrial integrity. Mitochondrial homeostasis is a broad term encompassing all aspects of maintenance of mitochondrial integrity.

MITOCHONDRIAL MEMBRANE INTEGRITY

Mitochondrial is a double membrane-bound organelle. Outer mitochondrial membrane (OMM) is separated from the inner mitochondrial membrane (IMM) by intermembranous space (IMS). Although OMM is a highly permeable structure, IMM is impermeable to most molecules. It contains specialized transporters and determines the type of molecules that can enter mitochondria. Moreover, it is highly convoluted as cristae that harbor electron transport chain. IMM is impermeable to protons and thus maintains the proton gradient that is crucial for ATP generation. IMM is enriched in a unique phospholipid, cardiolipin. Cardiolipin promotes the proper organization of respiratory chain complexes in a supercomplex and is necessary for cristae formation. It is synthesized from phosphatidic acid (PA) in mitochondria and remodeled after synthesis to acquire unique characteristics. Defects of cardiolipin synthesis and remodeling are outlined in Table 14.1.

MITOCHONDRIAL QUALITY CONTROL

Mitochondria contain about 1500 proteins required for its many essential cellular functions. Maintenance of mitochondrial proteome is crucial for its optimal function. Hence, there are mechanisms in place to ensure the maintenance of healthy mitochondrial proteome and removal of abnormal and damaged proteins. In addition, mitochondria are dynamic organelles and undergo fission and fusion. Mitochondrial fusion helps dilution of damaged proteins and maintains the overall efficacy of mitochondria. Mitochondrial fission helps to segregate the damaged proteins so that it can be removed by autophagy (mitophagy). When these control measures fail, overwhelming mitochondrial damage triggers cell death by initiating apoptosis. Hence, mitochondrial quality control measures are operative

Mitochondrial Medicine. https://doi.org/10.1016/B978-0-12-817006-9.00014-9

TABLE 14.1
Diseases of Mitochondrial Membrane Integrity

Mechanism of Deficiency	Gene	Clinical Presentation
Cardiolipin synthesis	*AGK* (*AGK* codes for acylglycerol kinase that phosphorylates diacylglycerol to phosphatidic acid, a precursor of cardiolipin. In addition, it also participates in the import of mitochondrial membrane proteins)	Sengers syndrome: congenital cataract, hypertrophic cardiomyopathy, skeletal myopathy, exercise intolerance, lactic acidosis, and increased urinary 3-methyl glutaconic acid (3-MGA).
Cardiolipin remodeling	*TAZ* (*TAZ* codes for Tafazzin that remodels cardiolipin necessary for its optimal function)	Barth syndrome: cardiomyopathy, skeletal myopathy, neutropenia, growth retardation, 3-MGA.
	DNAJC19 (DNAJC19 protein has a dual role in cardiolipin remodeling and mitochondrial protein import)	Dilated cardiomyopathy with ataxia (DCMA) syndrome: dilated cardiomyopathy, nonprogressive cerebellar ataxia, growth failure, testicular agenesis, 3-MGA. Additionally, microcytic anemia, intellectual disability, optic atrophy, and hepatic steatosis can be seen.

at the molecular level (mitochondrial proteome homeostasis), organellar level (mitochondrial fission and fusion), and cellular level (mitophagy).

Mitochondrial quality control at the molecular level: Most of the mitochondrial proteins are encoded by nuclear genes and imported in the mitochondria. Proteins targeted to mitochondria are imported across the OMM by the translocase of the outer membrane (TOM). The proteins destined for the mitochondrial matrix are attached with a mitochondrial targeting signal known as presequence and bind to translocase of inner membrane 23 (TIM23) for translocation to the matrix. Some of the proteins destined to IMM also use the same pathway. Other proteins of IMM, OMM proteins, and proteins of IMS utilize different biogenesis pathways. After import into the matrix, presequence is cleaved and the proteins undergo folding and assembly. Mitochondria matrix chaperones such as HSP60 and HSP70 assist in protein folding. Protein misfolding/unfolding due to mutations, oxidative damage, or other mechanisms lead to activation of mitochondrial matrix proteases such as LON and CLPXP. Misfolded proteins are directly degraded by LON, while CLPXP degradation generates peptides that activate transcription of nuclear genes for mitochondrial chaperones. This is called the mitochondrial unfolded protein response (UPR). Protein quality control in IMM is primarily achieved by the AAA class of proteases. They are classified as matrix facing (m-AAA) and IMS facing (i-AAA). Damaged, misfolded, or unfolded OMM proteins are ubiquitinated and removed by the ubiquitin-proteasome system. Table 14.2 summarizes the diseases associated with genes of mitochondrial protein quality control.

Mitochondrial quality control at the organellar level: Mitochondria are dynamic organelles. They undergo fission and fusion to enable distribution inside the cell. In addition, mitochondrial fusion provides an additional layer of defense to maintain quality. The damaged proteins get diluted in fused elongated mitochondria, and hence overall function of mitochondria remains preserved. Fusion of OMM is mediated by mitofusins, MFN1 and MFN2. Fusion of IMM is mediated by the protein, Optic atrophy 1 (OPA1). Failure of mitochondrial fusion results in fragmented dysfunctional mitochondria. Table 14.3 outlines the mitochondrial diseases of defects in fusion.

Mitochondrial quality control at the cellular level: When protein damage is overwhelming, mitochondrial fission helps to segregate the damaged mitochondrial portion from the remaining mitochondria so that it can be removed by a specialized form of autophagy called "mitophagy." Mitochondrial fission is mediated by a dynamin-related protein, DRP1. Mitophagy is mediated by two important molecules, PARKIN and

TABLE 14.2
Diseases of Mitochondrial Protein Quality Control

Gene	Pathophysiology	Clinical Presentation
HSPD1	HSPDI encodes for HSP60 that is a mitochondrial matrix chaperone	Spastic paraplegia type 13 (autosomal dominant) Hypomyelinating leukodystrophy (autosomal recessive)
CLPP	CLPP encodes for a subunit of mitochondrial matrix protease CLPXP	Perrault syndrome (sensorineural hearing loss, ovarian failure) type 3
SPG7	SPG7 encodes for paraplegin that is a subunit of m-AAA	Spastic paraplegia type 7
AFG3L2	AFG3L2 encodes for a catalytic subunit of m-AAA	Spinocerebellar ataxia type 28 (autosomal dominant) Spastic ataxia type 5 (autosomal recessive)

TABLE 14.3
Diseases of Impaired Mitochondrial Fusion

Mechanism of Deficiency	Gene	Clinical Presentation
Impaired fusion of OMM	MFN2	Charcot-Marie-Tooth (CMT) disease type 2A: progressive distal motor weakness and sensory loss. Both autosomal dominant and recessive forms are described.
Impaired fusion of IMM	OPA1	Autosomal dominant optic atrophy

PINK1. In healthy mitochondria, PINK1 is imported into the IMM and then cleaved. In damaged mitochondria, this process is inhibited; hence PINK1 starts accumulating on the OMM where they recruit PARKIN from the cytoplasm. PARKIN is a ubiquitin ligase. It ubiquitinates OMM proteins signaling the mitophagy process. Impaired mitophagy leads to accumulation of damaged, dysfunctional mitochondria. Mutations in *PINK1* and *PARKIN* genes are associated with familial Parkinson's disease.

GENETICS

Disorders of mitochondrial homeostasis due to nuclear gene mutations are inherited in autosomal recessive, autosomal dominant, and X-linked manner. Some genes are implicated in both autosomal dominant and recessive disorders (Table 14.4).

DIAGNOSIS

Diagnosis can be made by targeted genetic testing if the clinical picture is characteristic. Different molecular genetic diagnostic approaches are summarized in Table 14.5.

Blood lactate may not be elevated. Urine organic acid may show 3-MGA in disorders of cardiolipin synthesis and remodeling such as Barth syndrome and Sengers syndrome.

MANAGEMENT

A diagnosis of a disorder of mitochondrial homeostasis should lead to an evaluation for other systemic manifestations known to be associated with the condition. Management is mainly symptomatic, avoidance of mitotoxic medications, physical and occupational therapy (Table 14.6). Mitochondrial supplements such as Coenzyme Q10, L-carnitine, and Riboflavin can be tried. For research trials, "clinicaltrials.gov" is a very useful resource. Periodic evaluations for cardiac, endocrine, ophthalmologic, and audiological complications are needed and should be tailored according to the specific etiology. A multidisciplinary approach (audiologist, neurologist, ophthalmologist, endocrinologist, cardiologist, physical therapy and rehabilitation specialist) is needed on an individual case basis for optimal management.

TABLE 14.4
Inheritance Pattern of Diseases of Mitochondrial Homeostasis

Inheritance Pattern	Gene	Clinical Phenotype
X-linked	*TAZ*	Barth syndrome
Autosomal dominant	*HSPD1*	Spastic paraplegia type 13
	AFG3L2	Spinocerebellar ataxia type 28
	MFN2	CMT2A2A
	OPA1	Autosomal dominant optic atrophy.
Autosomal recessive	*HSPD1*	Hypomyelinating leukodystrophy
	AFG3L2	Spastic ataxia type 5
	MFN2	CMT2A2B
	Most of the other genes	See text

TABLE 14.5
Molecular Genetic Diagnostic Approaches

Molecular Genetic Test	Remarks
Targeted nuclear gene sequencing	*TAZ*: cardiomyopathy, neutropenia, myopathy, 3-MGA, X-linked inheritance
Targeted gene panel	For example, if CMT is diagnosed clinically, CMT gene panel can be ordered.
Combined mitochondrial genome and nuclear gene panel	A popular approach is to combine mtDNA sequencing with the sequencing of known nuclear genes related with mitochondrial function. If family history is suggestive of autosomal inheritance, only nuclear gene panel can be ordered first.
Whole exome/genome sequencing with mitochondrial genome sequencing	This is the most comprehensive approach. It not only allows diagnosis of a mitochondrial disorder but also other disorder that is in the differential diagnosis. However, there is also an increased likelihood of finding variants of unknown significance.

TABLE 14.6
Systemic Manifestations of Mitochondrial Homeostasis Diseases and Their Management

Organ System	Manifestations	Management
Constitutional	Failure to thrive	Optimize nutrition, gastrostomy tube feeding
Ophthalmology	Optic neuropathy	Visual rehabilitation
Audiology	Sensorineural deafness	Cochlear implant, hearing aids, sign language
Neurology	Intellectual disability, developmental delays, hypotonia, spasticity, ataxia, seizures	Anticonvulsant, physical therapy
Musculoskeletal	Myopathy, contractures	Physical therapy, occupational therapy
Endocrinology	Ovarian failure	Hormone replacement, assisted reproduction, etc.
Cardiovascular	Cardiomyopathy	Angiotensin-converting enzyme inhibitors, heart transplant
Respiratory	Respiratory muscle weakness	Ventilator support
Gastrointestinal	Dysphagia, gastroparesis	Motility agents

SUGGESTED READING

1. Moehle EA, Shen K, Dillin A. Mitochondrial proteostasis in the context of cellular and organismal health and aging. *J Biol Chem*. 2018. jbc.TM117.000893. [Epub ahead of print].

2. Quirós PM, Langer T, López-Otín C. New roles for mitochondrial proteases in health, ageing and disease. *Nat Rev Mol Cell Biol*. 2015;16:345–359.

3. Karbowski M, Neutzner A. Neurodegeneration as a consequence of failed mitochondrial maintenance. *Acta Neuropathol*. 2012;123:157–171.

4. Lu YW, Claypool SM. Disorders of phospholipid metabolism: an emerging class of mitochondrial disease due to defects in nuclear genes. *Front Genet*. 2015;6:3.

Disorders of Pyruvate Metabolism and Tricarboxylic Acid Cycle

ABSTRACT

Pyruvate is derived from glucose by glycolysis in the cytoplasm. After entering mitochondria, it is converted to acetyl coenzyme A, which enters the tricarboxylic acid (TCA) cycle to generate reducing equivalents NADH and FADH2. Although not strictly respiratory chain disorders, the disorders of pyruvate metabolism and TCA cycle closely mimic respiratory chain defects as the TCA cycle is the main electron donor of the respiratory chain. Hence, they are often discussed with the diseases of the respiratory chain.

KEYWORDS

Krebs cycle; Pyruvate carboxylase; Pyruvate dehydrogenase; Pyruvate transporter; Tricarboxylic acid cycle.

BACKGROUND

Carbohydrate and fat are the main energy source of the body. Glucose and other monosaccharides are converted to pyruvate via glycolysis pathway in the cytoplasm. Pyruvate is then transported across the inner mitochondrial membrane by pyruvate transporter. Once inside mitochondria, the pyruvate dehydrogenase enzyme converts pyruvate to acetyl CoA. Fatty acid oxidation also generates acetyl CoA as the end product. Acetyl CoA produced from carbohydrate and fat metabolism then enters the tricarboxylic acid (TCA) cycle by combining oxaloacetate to form citrate (Fig. 15.1). The TCA cycle generates reducing equivalents NADH and FADH$_2$, which then pass electrons to the electron transport chain (ETC)-generating ATP. Diseases of pyruvate metabolism affect energy production from carbohydrates, while TCA cycle disorders affect energy production from both carbohydrate and fat. Disorders of pyruvate metabolism and TCA cycle results in the failure of energy production and mimic respiratory chain disorders.

DISORDERS OF PYRUVATE METABOLISM

Pyruvate is derived from glucose and other monosaccharides via glycolysis pathway. In the aerobic condition, pyruvate is transported into the mitochondria where it is converted to acetyl CoA by pyruvate dehydrogenase enzyme complex. In anaerobic condition, pyruvate is reduced to lactate by the enzyme lactate dehydrogenase. Pyruvate can interconvert to alanine. Pyruvate is also an important molecule in the gluconeogenesis pathway. It is carboxylated to oxaloacetate by pyruvate carboxylase enzyme. Oxaloacetate is a tricarboxylic cycle intermediate. Oxaloacetate is decarboxylated by phosphoenolpyruvate carboxykinase enzyme to form phosphoenolpyruvate that then follows the reverse of glycolysis pathway to generate glucose. Thus, pyruvate is an important molecule in energy metabolism as well as gluconeogenesis. Pyruvate metabolism defects primarily affect the brain. The main disorders of pyruvate metabolism are summarized in (Table 15.1).

DISORDERS OF TRICARBOXYLIC ACID CYCLE

Tricarboxylic acid (TCA) cycle is the main electron donor of the ETC. In addition, succinate dehydrogenase (SDH), which catalyzes the conversion of succinate to fumarate in the TCA cycle, is also part of ETC complex II (Fig. 15.2). Disorders of TCA cycle mimic OXPHOS defects. Table 15.2 enlists inborn errors of TCA cycle. Some TCA cycle defects are also implicated in cancer pathogenesis as accumulated TCA cycle intermediates act as oncometabolites. Germline mutations in genes encoding for TCA cycle enzymes have been associated with inherited cancer predispositions (Table 15.3).

Mitochondrial Medicine. https://doi.org/10.1016/B978-0-12-817006-9.00015-0

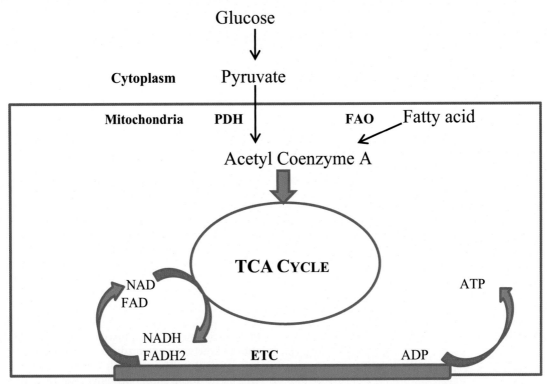

FIG. 15.1 Schematic representation of pyruvate metabolism and TCA cycle in relation to ETC. ADP: adenosine diphosphate; ATP: adenosine triphosphate; ETC: electron transport chain; FAD: flavin adenine dinucleotide; FAO: fatty acid oxidation; NAD: nicotinamide adenine dinucleotide; PDH: pyruvate dehydrogenase; TCA: tricarboxylic acid.

TABLE 15.1
Disorders of Pyruvate Metabolism

Disease	Pathophysiology	Gene	Clinical Presentation
Pyruvate dehydrogenase (PDH) deficiency	In PDH deficiency, the conversion of pyruvate to acetyl CoA is impaired leading to lactic acidosis and carbohydrate intolerance	PDH is a multiplex enzyme made up of three units: E1, E2, and E3. E1 has two components E1α and E1β. Mutations in *PDHA1* encoding E1α subunit is the most common cause of PDH deficiency and is inherited in X-linked dominant manner.	Varies in severity. **Neonatal:** congenital lactic acidosis, hypotonia, encephalopathy, brain malformations (agenesis of corpus callosum). **Infantile:** global developmental delay, seizures, Leigh syndrome-like presentation **Late onset:** Recurrent ataxia precipitated by carbohydrate-rich meal, peripheral neuropathy.

TABLE 15.1
Disorders of Pyruvate Metabolism—cont'd

Disease	Pathophysiology	Gene	Clinical Presentation
Pyruvate carboxylase (PC) deficiency	In PC deficiency, the conversion of pyruvate to oxaloacetate is impaired. This leads to impairment of the TCA cycle as well of gluconeogenesis. The acetyl CoA produced from fatty acid oxidation is not channeled adequately through TCA cycle due to deficient oxaloacetate leading to ketosis.	*PC*	PC deficiency leads to lactic acidosis, hypoglycemia, ketosis, and energy failure. The presentation varies in severity. **Neonatal (Type B):** Present shortly after birth with encephalopathy, lactic acidosis, hypoglycemia, ketosis, hypotonia, and abnormal neuroimaging findings such as periventricular cysts, ischemic-like lesions, or ventricular dilation. **Infantile (Type A):** Present between 2 and 5 months of age with developmental delays and hypotonia. Vomiting, tachypnea, and lactic acidosis are precipitated by intercurrent illnesses. Neuroimaging may show cortical atrophy, hypomyelination, and periventricular cysts.
Pyruvate transporter deficiency	Transport of pyruvate across the mitochondrial membrane is impaired	Mitochondrial pyruvate carrier is made up of two units (MPC1 and MPC2) Mutations in the *MPC1* gene have been reported in patients with pyruvate transporter deficiency	Delayed development, progressive microcephaly, neuroregression, hypotonia, lactic acidosis, and failure to thrive. Neuroimaging may show cortical atrophy, periventricular leukomalacia, and calcifications.

GENETICS

Mitochondrial diseases due to defects in pyruvate metabolism and TCA cycle caused by nuclear gene mutations are usually inherited in an autosomal recessive fashion. However, autosomal dominant and X-linked inheritances are also found (Table 15.4). Occasionally, the condition is sporadic due to de novo mutation.

DIAGNOSIS

Diagnosis can be suspected based on clinical and biochemical findings. For confirmation, molecular genetic testing or enzyme assay is needed. Different diagnostic tests are summarized in Table 15.5.

MANAGEMENT

Once a specific diagnosis is established, evaluation for other systemic manifestations associated with the condition should be done. Management is mainly symptomatic. For some conditions, supplementation of cofactors may be helpful (thiamine for PDH deficiency, biotin for PC deficiency). The ketogenic diet is beneficial in PDH deficiency as it provides acetyl CoA which can enter the TCA cycle by combining with oxaloacetate. In PC deficiency, there is a deficiency of TCA intermediates. Hence, measures to replenish TCA cycle intermediates (anaplerosis), such as citrate therapy and triheptanoin supplementation, are helpful.

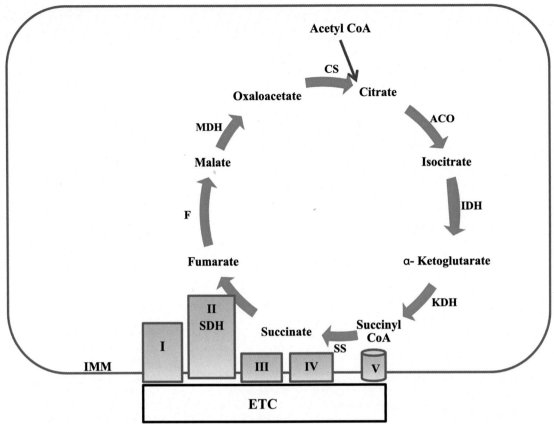

FIG. 15.2 Diagrammatic representation of the tricarboxylic acid cycle and its relation to the electron transport chain. ACO: aconitase; CS: citrate synthase; ETC: electron transport chain; F: fumarase; IDH: isocitrate dehydrogenase; IMM: inner mitochondrial membrane; KDH: α-ketoglutarate dehydrogenase; MDH: malate dehydrogenase; SDH: succinate dehydrogenase; SS: succinyl CoA synthetase. A molecule of NAD is converted to NADH at IDH, KDH, and MDH reactions that donate electrons to complex I. A molecule of FAD is converted to FADH2 by the SDH reaction that donates electrons to complex II. A molecule of GDP is converted to GTP by the SS reaction.

TABLE 15.2
Inherited Disorders of Tricarboxylic Acid Cycle

Disease	Pathophysiology	Gene	Clinical Presentation
α-Ketoglutarate dehydrogenase (KDH) deficiency	In KDH deficiency, the conversion of α-ketoglutarate to succinyl CoA is impaired	KDH is a multiplex enzyme made up of three units: E1, E2, and E3. E1 and E2 are specific for KDH. E3 component is the same as in PDH. E1 is encoded by *OGDH*.	Patients present in infantile or early childhood period. They present with global delays, hypotonia, ataxia, and seizures. There are lactic acidosis and high α-ketoglutaric acid in urine.

TABLE 15.2
Inherited Disorders of Tricarboxylic Acid Cycle—cont'd

Disease	Pathophysiology	Gene	Clinical Presentation
Succinyl CoA synthetase (SS)/ Succinyl CoA ligase deficiency	In SS deficiency, the conversion of succinyl CoA to succinate is impaired. In addition, there is an accumulation of methylmalonic acid due to inhibition of the conversion of methylmalonyl CoA to succinyl CoA. SS deficiency also causes disruption of mitochondrial nucleoside diphosphate kinase activity resulting in mtDNA depletion.	SS is made up of α and β subunits. *SUCLG1* and *SUCLA2* code for α and β subunits, respectively.	Infantile lactic acidosis, encephalomyopathy, hypotonia, methylmalonic aciduria. Brain MRI may show lesions seen in Leigh syndrome. Mitochondrial DNA depletion can be estimated from the muscle biopsy specimen.
Succinate dehydrogenase (SDH) deficiency	SDH catalyzes the conversion of succinate to fumarate. It is made up of four subunits: A, B, C, and D. Subunits A and B are also part of complex II. FAD is converted to FADH2 by the SDH reaction. An electron from FADH2 is transferred to Coenzyme Q by complex II.	*SDHA, SDHB, SDHC*, and *SDHD* code for A, B, C, and D subunits, respectively.	SDH deficiency presents with encephalomyopathy resembling Leigh syndrome. In addition, the presentation could be cardiomyopathy, skeletal myopathy, exercise intolerance, ataxia, or optic atrophy.
Fumarase deficiency	In fumarase deficiency, the conversion of fumarate to malate is impaired.	*FH*	Most patients present in infancy with global developmental delays, hypotonia, and seizures. In addition, dysmorphic features consisting of frontal bossing, depressed nasal bridge, and hypertelorism may be seen. Neuroimaging may show agenesis of corpus callosum, hydrocephalus, cortical malformations, and choroid plexus cysts.

TABLE 15.3
Inherited Cancer Predisposition Syndromes and Tricarboxylic Acid Cycle

Cancer Predisposition Syndrome	TCA Cycle Enzyme	Gene	Pathophysiology
Hereditary paraganglioma and pheochromocytomas	Succinate dehydrogenase (SDH)	*SDHA, SDHB, SDHC, SDHD, SDHAF2. SDHA, SDHB, SDHC*, and *SDHD* code for A, B, C, and D subunits of SDH, respectively. *SDHAF2* code for complex II assembly factor.	Heterozygous germline mutations in SDH genes are inherited in autosomal dominant fashion. Loss of heterozygosity leads to complete loss of protein in the affected cells. The accumulation of succinate leads to stabilization of hypoxia-inducible transcription factor-1α (HIF-1α) leading to metabolic reprogramming of cell that promotes carcinogenesis.

(continued)

TABLE 15.3
Inherited Cancer Predisposition Syndromes and Tricarboxylic Acid Cycle—cont'd

Cancer Predisposition Syndrome	TCA Cycle Enzyme	Gene	Pathophysiology
Multiple cutaneous and uterine leiomyomas (MCUL) Hereditary leiomyomatosis and renal cell cancer (HLRCC)	Fumarase (Fumarate hydratase)	FH	Heterozygous germline mutation in FH is inherited in autosomal dominant fashion. Loss of heterozygosity leads to complete loss of protein in the affected cells. The accumulation of fumarate leads to stabilization of hypoxia-inducible transcription factor-1α (HIF-1α) leading to metabolic reprogramming of cell that promotes carcinogenesis

TABLE 15.4
Inheritance Pattern of the Disorders of Pyruvate Metabolism and Tricarboxylic Acid Cycle

Inheritance Pattern	Gene	Clinical Phenotype
X-linked	PDHA1	Pyruvate dehydrogenase deficiency
Autosomal dominant	SDHA SDHB SDHC SDHD SDHAF2	Hereditary paraganglioma and pheochromocytomas
	FH	Multiple cutaneous and uterine leiomyomas (MCUL) Hereditary leiomyomatosis and renal cell cancer (HLRCC)
Autosomal recessive	Most of the other genes	See text

TABLE 15.5
Diagnostic Tests in Disorders of Pyruvate Metabolism and Tricarboxylic Acid Cycle

Diagnostic Test	Remarks
Blood glucose	Pyruvate carboxylase (PC) deficiency may present as hypoglycemia as PC is a gluconeogenesis pathway enzyme.
Blood lactate and pyruvate	High lactate. Lactate to pyruvate ratio is normal (10–20) in pyruvate dehydrogenase (PDH) deficiency.
Blood ketones	There is increased β-hydroxybutyrate and acetoacetate in blood in PC deficiency as the acetyl CoA produced from fat oxidation are not completely utilized by the TCA cycle and are directed toward ketogenesis.

TABLE 15.5
Diagnostic Tests in Disorders of Pyruvate Metabolism and Tricarboxylic Acid Cycle—cont'd

Diagnostic Test	Remarks
Urine studies	Urine organic acid shows ketonuria, high lactate, high pyruvate, and decreased amount of TCA cycle intermediates in PC deficiency. In defects of TCA cycle, TCA cycle intermediates above the block accumulate and are elevated in urine. For example, high fumaric acid in fumarase deficiency.
Neuroimaging	Brian imaging may show agenesis of the corpus callosum in PDH and fumarase deficiencies. In PC, periventricular cysts are often found. Leigh syndrome-like lesions can be seen in SDH or succinyl CoA ligase deficiencies.
Molecular genetic test	Targeted gene sequencing if the clinical and biochemical profile is characteristic. Gene panel testing or WES otherwise.
Enzyme studies	Assay of PC and PDH activities can be done from skin fibroblasts or peripheral blood lymphocytes.

SUGGESTED READING

1. Gray LR, Tompkins SC, Taylor EB. Regulation of pyruvate metabolism and human disease. *Cell Mol Life Sci.* 2014;71: 2577−2604.
2. Raimundo N, Baysal BE, Shadel GS. Revisiting the TCA cycle: signaling to tumor formation. *Trends Mol Med.* 2011;17: 641−649.
3. Brière JJ, Favier J, Gimenez-Roqueplo AP, Rustin P. Tricarboxylic acid cycle dysfunction as a cause of human diseases and tumor formation. *Am J Physiol Cell Physiol.* 2006;291: C1114−C1120.
4. Rustin P, Munnich A, Rötig A. Succinate dehydrogenase and human diseases: new insights into a well-known enzyme. *Eur J Hum Genet.* 2002;10:289−291.

CHAPTER 15.1

Pyruvate Dehydrogenase Deficiency

PYRUVATE DEHYDROGENASE COMPLEX

Pyruvate dehydrogenase (PDH) enzyme is a complex made of three components: E1, E2, and E3 (Fig. 15.1.1). E1 is made of two subunits: E1α and E1β. The E1 unit catalyzes decarboxylation of pyruvate in the presence of thiamine pyrophosphate (TPP) to form hydroxyethyl thiamine pyrophosphate (HETPP). E1 catalyzes the rate-limiting step in PDH complex. It is activated by PDH phosphatase and deactivated by PDH kinase. E2 catalyzes the next step in the conversion of pyruvate to acetyl CoA by its dihydrolipoamide acetyltransferase (DLAT) activity. It transfers the hydroxyethyl group from HETPP to E2 bound oxidized lipoamide-generating acetyl lipoamide. Acetyl group from acetyl lipoamide is then transferred to coenzyme A to generate acetyl CoA. In this process, a reduced form of lipoamide, dihydrolipoamide is generated. Finally, E3 unit catalyzes the oxidation of dihydrolipoamide to lipoamide. Dihydrolipoamide dehydrogenase (DLD) activity of E3 is not unique to PDH as it also participates in α-ketoglutarate dehydrogenase, branch chain keto acid dehydrogenase, and glycine cleavage complexes. E3 is bound to E1 and E2 core by E3-binding protein.

CLINICAL PRESENTATION

In pyruvate dehydrogenase (PDH) deficiency, the conversion of pyruvate to acetyl CoA is impaired. As acetyl CoA is the end product of glucose and other monosaccharides to enter TCA cycle for ATP generation, there is a deficiency of energy production from

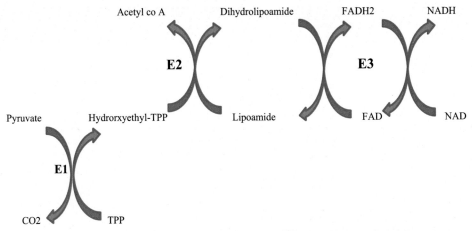

FIG. 15.1.1 Pyruvate dehydrogenase complex. TPP: thiamine pyrophosphate.

carbohydrates. Hence, tissues such as the brain that primarily rely on glucose are affected in PDH deficiency. Clinical presentation varies in severity. Neonatal form presents with severe lactic acidosis, tachypnea, hypotonia, and encephalopathy. Brain imaging may show agenesis of the corpus callosum, subependymal cysts, ventriculomegaly, and cortical atrophy. These patients usually die in the neonatal period. Patients may present later in infancy with global developmental delays, hypotonia, seizures, breathing abnormalities, microcephaly, and encephalopathy. Brain imaging may show agenesis or dysgenesis of corpus callosum, or other abnormalities mentioned earlier. Neuroimaging findings sometimes resemble Leigh syndrome showing bilaterally symmetrical lesions in the brain stem and basal ganglia. Facial dysmorphism consisting of frontal bossing, long philtrum, upturned nose, and wide nasal bridge resembling fetal alcohol syndrome may be seen. Rare late-onset cases may present with recurrent ataxia precipitated by carbohydrate-rich meal.

Patients with E3 deficiency (DLD deficiency) may present differently as E3 is also part of other enzyme complexes. Two main presentations have been described: early-onset neurological presentation, and isolated hepatic presentation. Patients with early-onset neurological type present with encephalopathy, metabolic acidosis, lactic acidosis, hypotonia, and encephalopathy in infancy. They frequently succumb to the metabolic decompensations. Patients with isolated hepatic type present at any age with recurrent liver failure. In between liver failure episodes, patients are asymptomatic. However, acute liver failure episodes can be fatal.

GENETICS

PDH deficiency is genetically heterogeneous. Mutations in genes encoding E1α, E1β, E2, E3, E3-binding protein, cofactors (TPP, lipoamide), and regulators (PDH phosphatase, PDH kinase) may cause PDH deficiency (Table 15.1.1). *PDHA1* gene encoding E1α is the most common cause of PDH deficiency (approximately 60%) and is inherited in X-linked semidominant fashion. Hence, males are more commonly and more severely affected with PDH deficiency. Females tend to have a more uniform presentation of varying severity dependent on variable lyonization. Males with *PDHA1* mutations tend to have missense mutation, while females usually have frameshift deletions or insertions resulting in premature termination. In about 25% cases, the mother of an affected child with *PDHA1* mutation is a carrier.

DIAGNOSIS

Diagnosis of PDH is suggested by the characteristic clinical and neuroradiological features. Blood lactate is usually high with normal (10−20) lactate to pyruvate ratio. In the presence of high lactate and characteristic clinical findings, molecular genetic studies are most commonly performed as the next step confirmatory diagnostic test. Table 15.1.2 enlists the diagnostic strategies for PDH deficiencies.

MANAGEMENT

Management is mainly symptomatic. Detailed neurological evaluation, anticonvulsants, and physical and occupational therapies are often needed. The ketogenic

TABLE 15.1.1
Genetics of PDH Deficiency

Gene	Inheritance Pattern	Remarks
PDHA1	X-linked	Most common cause of PDH deficiency Encodes E1α subunit
PDHB	Autosomal recessive	Encodes E1β subunit
DLAT	Autosomal recessive	Encodes E2 subunit
DLD	Autosomal recessive	Encodes E3 subunit
PDHX	Autosomal recessive	Second most common cause of PDH deficiency Encodes E3-binding protein
PDP1	Autosomal recessive	Encodes PDH phosphatase that activates E1
TPK1	Autosomal recessive	Encodes thiamine kinase that synthesizes TPP, a cofactor of PDH. In addition to PDH, activities of α-ketoglutarate dehydrogenase and branch chain keto acid dehydrogenase are impaired
LIAS	Autosomal recessive	Encodes lipoic acid synthetase, a cofactor of PDH. In addition to PDH, activities of α-ketoglutarate dehydrogenase, branch chain keto acid dehydrogenase, and glycine cleavage system are impaired.
NFU1 BOLA3	Autosomal recessive	Involved in the synthesis of Iron–Sulfur clusters. Iron–Sulfur clusters are needed for lipoic acid biosynthesis. In addition, Iron–Sulfur clusters are also required for the assembly of ETC complexes I, II, and III. Hence the clinical and biochemical manifestations resemble OXPHOS deficiency.

TABLE 15.1.2
Diagnosis of PDH Deficiency

Diagnostic Test	Remarks
Blood lactate and pyruvate	High lactate with normal (10–20) lactate to pyruvate ratio
Plasma amino acids	High alanine. In DLD, TPP, or lipoid acid synthesis defects branch chain amino acids and/or glycine may be high.
Urine organic acid	Increased lactic acid, pyruvic acid. In DLD, TPP, or lipoid acid synthesis defects, α-keto or α-hydroxy branch chain organic acids, and α-ketoglutaric acid may be elevated.
Molecular genetic test	PDHA1 gene sequencing with deletion/duplication study. If PDHA1 study is normal, other genes can be sequenced together as a panel
Enzyme studies	Activity of PDH complex or its components can be measured from skin fibroblasts. Immunohistochemistry studies may be done to evaluate E3-binding protein. PDH activity can also be measured in peripheral blood lymphocytes.

diet is the cornerstone of therapy. Acetyl CoA derived from fats can enter the TCA cycle and generate ATP. The ketogenic diet was shown to have a positive impact on clinical and biochemical parameters. Supplementation with thiamine should always be tried as TPP is a cofactor in PDH complex. Thiamine-responsive variant of PDH deficiency has been described. Dichloroacetate (DCA) is an inhibitor of PDH kinase and hence it facilitates the active state of PDH E1. DCA has been used in a dose of 25 mg/kg/day (two divided doses) by mouth. The comprehensive list of ongoing clinical trials for PDH deficiency can be found on clinicaltrails.gov.

SUGGESTED READING

1. Patel KP, O'Brien TW, Subramony SH, Shuster J, Stacpoole PW. The spectrum of pyruvate dehydrogenase complex deficiency: clinical, biochemical and genetic features in 371 patients. *Mol Genet Metabol.* 2012;106:385–394.
2. Shin HK, Grahame G, McCandless SE, Kerr DS, Bedoyan JK. Enzymatic testing sensitivity, variability and practical diagnostic algorithm for pyruvate dehydrogenase complex (PDC) deficiency. *Mol Genet Metabol.* 2017;122:61–66.
3. Sofou K, Dahlin M, Hallböök T, Lindefeldt M, Viggedal G, Darin N. Ketogenic diet in pyruvate dehydrogenase complex deficiency: short- and long-term outcomes. *J Inherit Metab Dis.* 2017;40:237–245.
4. Berendzen K, Theriaque DW, Shuster J, Stacpoole PW. Therapeutic potential of dichloroacetate for pyruvate dehydrogenase complex deficiency. *Mitochondrion.* 2006;6:126–135.

CHAPTER 15.2

Pyruvate Carboxylase Deficiency

PATHOGENESIS

Pyruvate carboxylase (PC) catalyzes carboxylation of pyruvate to oxaloacetate. Oxaloacetate is an intermediate of TCA cycle. Hence, PC deficiency results in depletion of TCA cycle intermediates and impairment of ATP production as the TCA cycle is the main electron donor for the respiratory chain. Oxaloacetate also participates in the gluconeogenesis pathway. It is converted to phosphoenolpyruvate (PEP) by phosphoenolpyruvate carboxykinase (PEPCK). PEP is then converted to glucose. Oxaloacetate combines with acetyl CoA produced from pyruvate dehydrogenase enzyme to form citrate, another TCA cycle intermediate. Citrate can also be exported to the cytoplasm where it is cleaved by citrate lyase to form acetyl CoA and oxaloacetate. Acetyl CoA produced in the cytoplasm is directed to fatty acid and cholesterol synthesis. Thus, PC is an important enzyme participating in energy metabolism, gluconeogenesis, and lipid synthesis (Fig. 15.3.1). In addition, there is urea cycle impairment. Aspartate derived from oxaloacetate binds to citrulline in the urea cycle to generate argininosuccinic acid. Deficiency of aspartate in PC deficiency leads to urea cycle impairment manifesting as hypercitrullinemia and hyperammonemia. Thus, PC deficiency can lead to energy failure, hypoglycemia, lactic acidosis, hyperammonemia, and ketosis (as acetyl CoA derived from fatty acid oxidation is not utilized completely via TCA cycle). Tissues dependent on high flux activity of TCA cycles such as the liver and brain are mainly affected in PC deficiency. In the brain, synthesis of neurotransmitters glutamate and γ-aminobutyric acid (GABA) are impaired as they are derived from α-ketoglutarate, another TCA intermediate.

CLINICAL PRESENTATION

Clinical presentation of PC deficiency varies in severity.

Neonatal form (Type B): Neonatal form is the most severe form. It manifests shortly after birth with tachypnea, encephalopathy, hypotonia, lactic acidosis, hypoglycemia, hyperammonemia, and ketosis. Additionally, hepatomegaly and seizures may be present. The prognosis is very poor, and most patients die within a few months.

Infantile form (Type A): Infantile form typically presents between the age of 2 and 5 months with failure to thrive, developmental delays, and hypotonia. Acute episodes of metabolic decompensation are precipitated by common illnesses manifesting as tachypnea, lethargy, and lactic acidosis. In addition, seizures and nystagmus may be present. This form is most common in North American Indians. Prognosis is poor as most patients die in infancy.

Benign form (Type C): This form is very rare presenting as acute episodes of lactic acidosis and ketosis precipitated by metabolic stressors such as infection and dehydration. These episodes respond rapidly to hydration and bicarbonate therapy.

GENETICS

PC deficiency is caused by homozygous or compound heterozygous mutations in the *PC* gene. Patients with Type A form usually have two missense mutations in homozygous or compound heterozygous state. Type B patients have at least one truncating mutation. PC is inherited in autosomal recessive fashion.

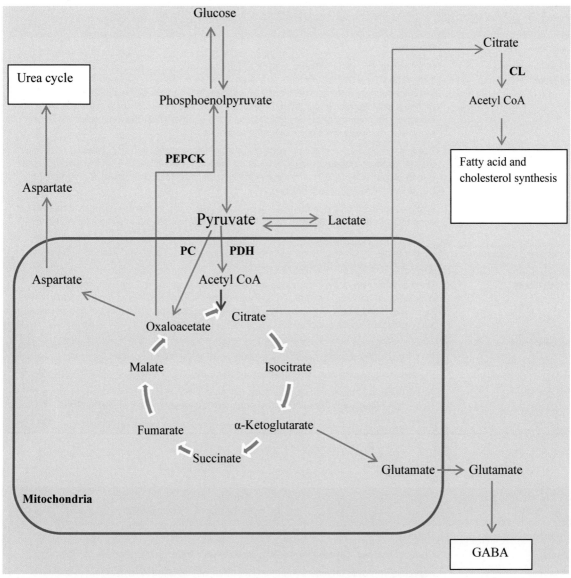

FIG. 15.2.1 Pyruvate carboxylase and metabolic pathways. CL: citrate lyase; GABA: γ-aminobutyric acid; OAA: oxaloacetate; PC: pyruvate carboxylase; PDH: pyruvate dehydrogenase; PEPCK: phosphoenolpyruvate carboxykinase.

DIAGNOSIS

Diagnosis of PC should be suspected in infants presenting with lactic acidosis, hypoglycemia, and ketosis. Type B patients present shortly after birth. They show most marked biochemical abnormalities. Type C patients may show biochemical abnormalities only during acute episodes. Table 15.2.1 enlists the diagnostic strategies and typical findings in PC deficiency.

MANAGEMENT

Management is largely symptomatic. Detailed neurological evaluation, anticonvulsants, and physical and occupational therapies are often needed. The main aim of treatment is to replenish the TCA cycle intermediates (anaplerosis). Triheptanoin has been used in PC deficiency as it is broken down to propionyl CoA, which is converted to succinyl CoA, a

TABLE 15.2.1
Diagnosis of PC Deficiency

Diagnostic Test	Remarks
Blood lactate and pyruvate	High lactate. In type B patients, lactate to pyruvate ratio is usually high (>25). However, in Type A and Type C, it is normal. There is an increase in reducing equivalent carrier malate in the cytoplasm as there is a deficiency of oxaloacetate and consequently aspartate in the mitochondrial matrix. Hence, there is reduced exchange of malate for aspartate across mitochondrial membrane by the malate-aspartate shuttle. Thus, the increase in NADH in cytoplasm favors conversion of pyruvate to lactate.
Blood ketones	There is increased β-hydroxybutyrate (H) and acetoacetate (A) in blood. However, H/A ratio may be decreased (particularly in type B) as there is reduced NADH in mitochondrial matrix due to impairment of the TCA cycle. Hence, there is a reduced conversion of acetoacetate to β-hydroxybutyrate that requires NADH.
Ammonia	High plasma ammonia due to impairment of the urea cycle as there is reduced availability of aspartate to combine with citrulline to form argininosuccinate, an intermediate of the urea cycle.
Plasma amino acids	High alanine, citrulline, and low glutamate and glutamine. In addition, there is high proline as lactic acidosis inhibits proline catabolism.
Urine studies	Urine studies may show bicarbonaturia as renal tubular acidosis is a complication of PC deficiency. In addition, generalized aminoaciduria may be seen. Urine organic acid shows ketonuria, high lactate, high pyruvate, and decreased amount of TCA cycle intermediates such as fumarate and succinate.
Neuroimaging	Typical findings in MRI of the brain are as follows: ventricular dilation, periventricular cysts, hypomyelination, subcortical leukodystrophy, and ischemia-like lesions.
Molecular genetic test	*PC* gene sequencing with deletion/duplication study.
Enzyme studies	Assay of PC activity in skin fibroblasts is diagnostic. However, it is not useful in terms of distinction among subtypes. PC activity can also be measured in peripheral blood lymphocytes.

TABLE 15.2.2
Treatment Strategies for PC Deficiency

Treatment Strategy	Rationale
Citrate	Corrects acidosis. Citrate therapy also replenishes citrate, an intermediate of TCA cycle.
Aspartate	Aspartate improves urea cycle function by combining citrulline to form argininosuccinic acid, a urea cycle intermediate that is deficient in PC deficiency due to inadequate aspartate
Triheptanoin	Replenishes succinyl CoA, a TCA cycle intermediate
Bitoin	Cofactor for PC enzyme
Liver transplantation	Improves biochemical profile and survival. May not change the neurological outcome.

TCA cycle intermediate. High-fat diet can aggravate ketosis while high carbohydrate diet aggravates lactic acidosis. A high protein, low fat, and low carbohydrate diet may be beneficial. Table 15.2.2 summarizes the current treatment strategies for PC deficiency. The comprehensive list of ongoing clinical trials for PC deficiency can be found on clinicaltrails.gov.

SUGGESTED READING

1. Wang D, De Vivo D. Pyruvate carboxylase deficiency [Updated 2018 Mar 1]. In: Adam MP, Ardinger HH, Pagon RA, et al., eds. *GeneReviews® [Internet]*. Seattle (WA): University of Washington, Seattle; June 2, 2009:1993−2018.

2. Marin-Valencia I, Roe CR, Pascual JM. Pyruvate carboxylase deficiency: mechanisms, mimics and anaplerosis. *Mol Genet Metabol.* 2010;101:9−17.

CHAPTER 16

Mitochondria and Aging

ABSTRACT

The role of mitochondria in aging and age-related disorders is an interesting area of research. Mitochondrial dysfunction is considered either directly or indirectly related to various aspects of aging. A better understanding of this connection may potentially open avenues to delay the aging process and/or enable healthy aging.

KEYWORDS

Aging; Longevity; Life span; Mitochondria; Reactive oxygen species; Telomere.

HUMAN AGING AND MITOCHONDRIA

Aging can be defined as an age-dependent progressive decline in physiological functions leading to a gradual decline in general well-being, reduction in mobility, and the onset of age-related diseases such as heart disease, type 2 diabetes, and Alzheimer's disease. Mitochondrial dysfunction is considered one of the hallmarks of aging. Mitochondrial function declines with age and contributes to many manifestations of the aging process.

Mitochondria are very important organelles as they perform vital functions such as ATP production, calcium homeostasis, and cell signaling. Mitochondria are the main source of reactive oxygen species (ROS), which is central to the original theory of aging and also a mediator of apoptosis. Hence, it is intuitive that mitochondria are integral to the aging process. In the following sections, the age-associated changes in mitochondria and its relation to the aging process will be explored.

MITOCHONDRIAL OXIDATIVE RESPIRATION AND AGING

With age, the capacity of mitochondrial respiratory chain to generate ATP declines. It is estimated that ATP production declines by an average of 8% every decade. Decreased efficiency of oxidative phosphorylation makes tissues vulnerable to damages from acute or chronic stresses with high ATP demand. The decline in ATP production in skeletal muscles leads to reduced physical activity and sarcopenia. It is also central to neuronal and cardiac myocyte injury and loss leading to age-related brain and heart pathologies. The decrease in respiratory chain function is due to age-related accumulation of mtDNA mutations leading to dysfunction of mtDNA-derived respiratory chain subunits. In addition, the expression of nuclear genes related to the respiratory chain is decreased. Telomere shortening which is another hallmark of aging is considered directly linked to decreased expression of nuclear genes. Telomere shortening is associated with reduced level of the peroxisome proliferator-activated receptor gamma coactivator—1 α and 1β (PGC-1α and 1β). The decrease in expression of these transcriptional activators leads to reduced mitochondrial biogenesis, mtDNA content, and oxidative phosphorylation.

Apart from accumulating mtDNA mutations and decline in mitochondrial biogenesis, other notable factors leading to decreased mitochondrial function are altered mitochondrial dynamics, increased ROS generation, and impaired mitochondrial proteostasis.

MITOCHONDRIAL DNA MUTATIONS AND AGING

Mitochondrial DNA is more prone to mutations than nuclear DNA. This is due to several reasons:

1. Mitochondrial DNA lacks histone that forms protective insulation and prevents damage to nuclear DNA.
2. Mitochondrial DNA is close to the site of ROS generation making them prone to injury by ROS.
3. The DNA repair mechanism of mitochondrial DNA is less efficient.

Mitochondrial Medicine. https://doi.org/10.1016/B978-0-12-817006-9.00016-2

4. Mitochondrial DNA replication error is caused by deficiencies in proofreading function of polymerase gamma (*POLG*). Evidence suggests that replication error is the most common mechanism of mutations seen in aging. The characteristic large-scale deletion and transition point mutations are due to replication error.

Mitochondrial DNA mutations can lead to decreased respiratory chain function and increased ROS generation and are important mediators of the aging process. The evidence that mtDNA mutations are related to the aging process comes from the "mutator" mice model. Mutator mice have POLG deficient in proofreading activity. Hence, they accumulate a large number of mtDNA mutations. These mice display phenotypes of premature aging such as hair loss, graying of hair, kyphosis, and weight loss. Their life span is shortened. However, the levels of mtDNA mutation seen with normal aging process are significantly less than that seen in the mutator mice and are well below the threshold level to cause a functional defect. Thus, the role of mtDNA mutations in the aging process is controversial. There are two potential mechanisms relating mtDNA mutations to aging:

1. There are hundreds to thousands of mitochondria per cell and mtDNA replicates independently of the cell division (relaxed replication). At each cell division, mitochondria segregate randomly between the daughter cells. Thus, the level of mtDNA mutation varies among different cells of a tissue and between different tissues. Hence, even when the total level of mutation is low, the mutation level may reach above threshold in some cells or tissues creating a mosaic pattern of oxidative phosphorylation deficiency. It also suggests that clonal expansion of mitochondrial DNA mutations that arise early in embryonic or postnatal life and undergo several replication cycles to reach threshold level is the main driving force behind aging.

2. The phenotype of premature aging in "mutator mice" appeared before the oxidative phosphorylation deficiency. This led to speculation that the somatic stem cell dysfunction caused by mtDNA mutations is the primary driving process of aging. Somatic stem cells are dependent on mitochondrial integrity for their regenerative function. Mitochondrial DNA mutations by increasing the level of ROS disrupt the function of these cells and drive the aging process.

MITOCHONDRIAL BIOGENESIS AND AGING

Apart from the decline in efficiency of the mitochondrial respiratory chain, mitochondrial mass also declines with age. As mentioned earlier, PGC-1α and β are the key regulators of mitochondrial biogenesis. With advancing age, the level of PGC-1α declines and has been associated with telomere shortening. In addition, poor physical activity seen with advanced age leads to a decline in PGC-1α level and subsequent decrease in mitochondrial mass. Physical activity has a beneficial effect on aging process by increasing PGC-1α level.

MITOCHONDRIAL DYNAMICS AND AGING

Mitochondrial are dynamic organelles and form reticular structures inside the cell. They undergo fission and fusion to maintain integrity. Through fusion, damaged mitochondria unite with healthy mitochondria and acquire undamaged genetic material. In addition, the accumulated toxic substances such as ROS can be diluted by this process. Mitochondrial fission leads to breakage of damaged and dysfunctional part of mitochondria from the rest so that it can be removed by a special form of autophagy called mitophagy. Thus, fusion and fission maintain mitochondrial functionality. Mitophagy tends to decline with age. This leads to the accumulation of damaged mitochondria and oxidative stress that may contribute to the aging process.

MITOCHONDRIAL ROS PRODUCTION AND AGING

ROS was considered central to the aging process in the original Harman theory of aging. It was postulated that excessive ROS leads to oxidation and degradation of nucleic acids, membrane lipids, and proteins resulting in aging. As mitochondria are the main source of ROS, this theory was later modified as the mitochondrial theory of aging. According to this theory, ROS leads to the damage of mitochondrial DNA due to its vicinity to the respiratory chain complexes. Mitochondrial DNA mutations cause dysfunction of the respiratory chain and further increase in the production of ROS thus creating a vicious cycle and "burst" in the ROS production. However, the pathogenesis of aging is more complex than this. Although oxidative stress increases with advancing age, it is not the main contributor to mitochondrial DNA damage. In addition, the mutator mice described earlier showed a very mild increase in ROS even in the presence of marked progeroid phenotype. On the contrary, the increased level of ROS has been found to increase the life span in various animal models by inducing an adaptive response that increases cellular resistance to stress. Moreover, ROS are important as signaling molecules in the cell cycle regulation. The level of ROS is tightly controlled in stem cells, and its dysregulation may cause aging by disrupting the integrity of somatic stem cells. Hence, the role of ROS is not as straight forward as initially postulated. A high ROS level is damaging and accelerates aging process while low ROS level modulates cellular adaptive response to stress in a positive way and is associated with longevity.

MITOCHONDRIAL PROTEOSTASIS AND AGING

Aging is associated with oxidative damage of proteins that leads to accumulation of misfolded proteins. The misfolded proteins trigger mitochondrial unfolded protein response. It consists of an increase in the production of chaperones and proteases. This response attempts to maintain mitochondrial proteostasis by enabling the correct folding of proteins or degrading the severely damaged proteins. Thus, mitochondrial proteostasis is important for maintaining mitochondrial integrity. Lon is an important mitochondrial protease. It degrades oxidized proteins and also has molecular chaperone function. It was shown that downregulation of Lon in human cells led to a decrease in mitochondrial proteolysis of damaged proteins and ultimately cell death. In addition, the level of Lon mRNA transcripts was lower in skeletal muscle of aging mice. Aging-related neurodegenerative diseases are associated with the accumulation of unfolded proteins such as amyloid in Alzheimer disease. Hence, abnormal proteostasis is related to the aging process.

MITOCHONDRIAL AND LONGEVITY

Mitochondria are not only important for energy production but also have vital signaling functions. Mitochondria communicate decline in their functional status to the nucleus. The adaptive response leads to the transcription of nuclear genes and stimulation of intracellular pathways that attempt to restore mitochondrial integrity and cellular homeostasis and increases cellular resistance to stress. This response is also called "mitohormesis" and increases cell survival and thus is important for longevity. Adaptive response to misfolded protein was described earlier. Similarly, loss of mitochondrial membrane potential, increase in ROS generation, and increase in certain metabolites

such as nicotinamide adenine dinucleotide (NAD+) and adenosine monophosphate (AMP) also activate this response. The adaptive response is not confined to the cell but distant tissues are also affected. Mitokines such as fibroblast growth factor 21 secreted in response to mitochondrial stress exert systemic beneficial effects including improved insulin sensitivity and weight loss.

Some important longevity pathways are as follows:
1. Insulin/insulin-like growth factor signaling (IIS) pathway: IIS pathway is an important biological pathway and regulates growth and cell proliferation. Downregulation of this pathway was found to increase life span in various species. Downregulation of this pathway leads to activation of the transcription factor FOXO that promotes mitochondrial biogenesis and increases expression of genes related with resistance to oxidative stress.
2. Mechanistic target of rapamycin (mTOR) pathway: mTOR pathway is very important and evolutionarily conserved as it senses environmental and intracellular growth and nutrient signals to coordinate many vital functions such as cell growth, proliferation, and inflammation. The inhibition of this pathway was found to be associated with increased life span in organisms. Although the mechanism of longevity mediated by mTOR inhibition is not clear, the overall reduction in mRNA translation enabling cells to efficiently handle damaged proteins and protein aggregates seen with aging is considered the likely explanation.
3. Sirtuin pathway: Sirtuins are deacetylase regulated by NAD + level. With advancing age NAD + level declines leading to decrease in Sirtuin. Sirtuin1 (encoded by *SIRT1*) is an important regulator of various intracellular processes. A decrease in SIRT1 expression leads to inhibition of PGC1α and FOXO resulting in decreased mitochondrial biogenesis and impaired antioxidative defense.
4. AMP-activated protein kinase (AMPK): AMPK acts as an energy sensor and is activated by an increase in AMP/ADP or AMP/ATP ratio. AMPK attempts to restore energy homeostasis by inhibiting energy-consuming anabolic pathways such as IIS and mTOR. In addition, it increases mitochondrial function by activation of PGC1α and FOXO resulting in increased mitochondrial biogenesis and antioxidative defense.

These pathways are interconnected and related to increased longevity seen in animal models with caloric restriction and exercise. Caloric restriction and exercise lead to an increase in AMP, NAD+, and mild elevation of ROS. Subsequently, AMPK and SIRT1 are activated.

AMPK and SIRT1 activation together lead to mitochondrial biogenesis via activation of PGC1α and resistance to oxidative stress via activation of FOXO. Caloric restriction also leads to inhibition of IIS and mTOR pathways. Inhibition of IIS causes FOXO activation. Inhibition of IIS and activation of AMPK and SIRT1 pathways also result in increased autophagy via FOXO activation. Autophagy is a survival mechanism enabling recycling of macromolecules and degradation of damaged organelles such as mitochondria. Thus overall effect is the promotion of mitochondrial biogenesis, improved proteostasis, and increased antioxidative defense resulting in longevity.

CONCLUSIONS

Mitochondrial dysfunction is an integral part of the aging process. However, some of the changes seen in mitochondria may be secondary to the aging process itself such as reduced physical activity and changed hormonal milieu. Although the role of mitochondria in the aging process is not as straightforward as postulated in the free radical theory, mitochondrial dysfunction plays a crucial role in the aging process. A better understanding of molecular mechanisms of mitochondrial dysfunction and its relation with aging will lead to interventions that may enable delaying the process of aging.

SUGGESTED READING

1. López-Otín C, Blasco MA, Partridge L, Serrano M, Kroemer G. The hallmarks of aging. *Cell.* 2013;153: 1194–1217.
2. Amigo I, da Cunha FM, Forni MF, et al. Mitochondrial form, function and signalling in aging. *Biochem J.* 2016;473: 3421–3449.
3. Bratic A, Larsson NG. The role of mitochondria in aging. *J Clin Invest.* 2013;123:951–957.
4. Ahlqvist KJ, Suomalainen A, Hämäläinen RH. Stem cells, mitochondria and aging. *Biochim Biophys Acta.* 2015;1847: 1380–1386.
5. Kauppila TES, Kauppila JHK, Larsson NG. Mammalian mitochondria and aging: an update. *Cell Metabol.* 2017;25: 57–71.
6. Raffaello A, Rizzuto R. Mitochondrial longevity pathways. *Biochim Biophys Acta.* 2011;1813:260–268.

Mitochondria, Obesity, Metabolic Syndrome, and Type 2 Diabetes

ABSTRACT

The role of mitochondria in late-onset metabolic problems such as obesity, metabolic syndrome, and type 2 diabetes is intriguing but far from completely understood. However, a better understanding of the role of mitochondria in the pathogenesis of these conditions will be very rewarding as it may lead to novel approaches to control these major public health problems.

KEYWORDS

Mitochondrial dysfunction; Metabolic syndrome; Obesity; Oxidative stress; Type 2 diabetes mellitus.

OBESITY, METABOLIC SYNDROME, AND TYPE 2 DIABETES

Obesity can be defined as the accumulation of excessive fat and more precisely as body mass index (BMI) > 30. Metabolic syndrome is co-occurrence of obesity, insulin resistance, atherogenic dyslipidemia (high triglyceride, low HDL cholesterol), and hypertension. Metabolic syndrome (MetS) increases the risk for type 2 diabetes mellitus (T2DM), myocardial infarction, stroke, and fatty liver. Diabetes mellitus (DM) is a metabolic disease characterized by chronic elevation of blood glucose. Type 1 DM is due to inadequate production of insulin by pancreatic β-cells while T2DM is because of insulin resistance. Although obesity, MetS, and T2DM are separate clinical entities, they are related and often occur together. Hence, their pathogeneses are interrelated.

These metabolic disorders are major public health problems worldwide and related to effects of urbanization—sedentary lifestyle and surplus nutrition. In a simplified model, excessive nutrition leads to lipid accumulation and obesity. The accumulated lipid, particularly diacylglycerol in the liver and skeletal muscle, inhibits insulin signaling causing insulin resistance. Insulin resistance leads to the state of relative hypoinsulinemia that leads to increased hormone-sensitive lipase activity and lipolysis in adipose tissue

resulting in increased plasma free fatty acid (FFA). This excessive FFA is shunted to the liver that results in excessive VLDL synthesis by liver resulting in hypertriglyceridemia. Excessive triglyceride in VLDL is exchanged for cholesterol in HDL resulting in low HDL cholesterol. Relative hypoinsulinemia leads to hyperglycemia. Pancreatic β-cells are overstimulated due to persistent hyperglycemia and finally decompensate leading to T2DM.

In addition, there is a chronic inflammation associated with these conditions. Obesity is associated with adipocyte hypertrophy and enlargement of adipose tissue resulting in local tissue hypoxia. This predisposes to necrosis of adipocytes and infiltration by macrophages. Macrophages secrete inflammatory mediators such as interleukin-6 (IL-6) and tumor necrosis factor α (TNF-α). Adipose tissue is also an endocrine organ and modulates metabolism by secreting adipokines. Hypertrophied adipocytes have disturbed adipokine secretory pattern. There is an increase in secretion of proinflammatory adipokines such as leptin and resistin, while a decrease in adiponectin. Adiponectin is antiinflammatory and increases insulin sensitivity. The imbalance between pro- and antiinflammatory adipokines leads to local and systemic inflammation. The overall effect of altered adipokine milieu and chronic systemic inflammatory status is increased food intake, insulin resistance, hepatic steatosis, endothelial dysfunction, and atherosclerosis.

ROLE OF MITOCHONDRIA

Oxidative stress is associated with obesity and plays a key role in the pathogenesis of MetS, insulin resistance, and T2DM. Accumulation of excessive FFA in adipose tissue causes activation of NADPH oxidase enzyme and generation of excessive reactive oxygen species (ROS) in adipocytes. The oxidative stress of adipocyte causes changes in their secretory pattern and more proinflammatory adipokines are secreted leading to infiltration of adipose tissue by macrophages and local inflammation. Persistent local inflammation leads to systemic inflammation and oxidative stress. Increased oxidative stress in skeletal muscles causes inhibition of insulin receptor signaling pathway and reduced transport of glucose in skeletal muscle leading to insulin resistance. Mitochondrial dysfunction contributes to this oxidative stress and hence plays a significant role in the pathogenesis of MetS and T2DM. It is postulated that nutrition overload will lead to mitochondrial overload of fatty acid and glucose in adipocytes leading to excessive acetyl coenzyme A, which results in an increase in NADH production from the tricarboxylic acid cycle. There is an increase of electron supply to the respiratory chain of mitochondria. Some of these high energy electrons spill off the chain generating ROS rather than ATP. Mitochondria are an important source of ROS, and ROS produced in physiological amount has an important role in intracellular signaling. However, the substrate overload in obesity leading to excessive ROS generation is harmful. Increased oxidative stress leads to a further decline in mitochondrial function, mitochondrial fission, and apoptosis of the adipocytes. Apoptosis of adipocytes triggers infiltration of adipose tissue by macrophages, the release of inflammatory mediators, and progression of systemic inflammation potentiating insulin resistance. Enhanced mitochondrial fission causes overproduction of ROS and is frequently associated with MetS and T2DM. Mitochondrial dysfunction also plays a significant role in pancreatic β-cell failure and progression from insulin resistance to T2DM. Increased metabolic demand on pancreatic β-cells due to chronic hyperglycemia leads to excessive ROS formation. As a compensatory mechanism, expression of uncoupling protein on inner mitochondrial protein is increased to dissipate proton gradient and sacrifice ATP production rate to lower ROS generation. However, the decrease in ATP generation also leads to a decrease in insulin secretion. When ROS generation is excessive and persistent, it leads to damage of mitochondrial membrane, release of cytochrome *c*, and apoptosis of β-cells resulting in reduced β-cell mass. Both these processes contribute to the development of T2DM.

However, the decline in mitochondrial function may also be secondary to insulin resistance. Insulin-resistant cells were shown to have reduced mitochondrial energy production and increased susceptibility to oxidative stress. In addition, the sedentary lifestyle associated with these conditions is associated with a decrease in mitochondrial protein content due to decreased mitochondrial biogenesis resulting in mitochondrial dysfunction.

CONCLUSIONS

Role of mitochondria in the pathogenesis of obesity, MetS, and T2DM is unknown. However, mitochondrial dysfunction either secondary to nutrition overload and excessive ROS generation, impaired biogenesis secondary to decreased physical activity, or as a consequence of insulin resistance itself contributes to the oxidative stress and systemic inflammation that is central in the pathogenesis of these disorders. Hence, interventions to increase mitochondrial biogenesis and function such as exercise or to reduce oxidative stress have roles in the treatment of these conditions.

SUGGESTED READING

1. Henriksen EJ, Diamond-Stanic MK, Marchionne EM. Oxidative stress and the etiology of insulin resistance and type 2 diabetes. *Free Radic Biol Med.* 2011;51:993−999.
2. Ma ZA, Zhao Z, Turk J. Mitochondrial dysfunction and β-cell failure in type 2 diabetes mellitus. *Exp Diabetes Res.* 2012;2012:703538.
3. Fex M, Nicholas LM, Vishnu N, et al. The pathogenetic role of β-cell mitochondria in type 2 diabetes. *J Endocrinol.* 2018; 236:R145−R159.
4. Burkart AM, Tan K, Warren L, et al. Insulin resistance in human iPS cells reduces mitochondrial size and function. *Sci Rep.* 2016;6:22788.

CHAPTER 18

Mitochondria and Nonalcoholic Fatty Liver Disease

ABSTRACT

Nonalcoholic fatty liver disease (NAFLD) is the most common chronic liver disease and a worldwide epidemic. The role of mitochondria in the pathogenesis of NAFLD is increasingly being recognized. Understanding the pathophysiology of NAFLD is very important to find novel ways to prevent and treat this global health problem.

KEYWORDS

Fatty liver; Hepatic steatosis; Mitochondrial dysfunction; Nonalcoholic fatty liver disease; Oxidative stress.

NONALCOHOLIC FATTY LIVER DISEASE

NAFLD involves a spectrum of conditions characterized by hepatic fat accumulation in the absence of secondary causes (significant alcohol intake, viral infection, etc.). The spectrum ranges from simple steatosis to nonalcoholic steatohepatitis (NASH) characterized by liver cell injury, inflammation, and fibrosis. Moreover, NASH can progress to cirrhosis and hepatocellular carcinoma.

NAFLD can be considered as a hepatic manifestation of metabolic syndrome (MetS). It is associated with obesity, insulin resistance, and type 2 diabetes mellitus (T2DM). Approximately, 70% of obese individuals with T2DM have NAFLD. In addition, NAFLD is an independent risk factor for cardiovascular disease. Hepatic fat accumulation simulates secretion of very low-density lipoprotein from the liver. There is a concomitant reduction in high-density lipoprotein. Thus, NAFLD increases risks of liver cirrhosis, hepatocellular carcinoma, and death from cardiovascular disease. Hence, understanding the pathogenesis of NAFLD is essential to decrease its prevalence and/or complications.

ROLE OF MITOCHONDRIA

NAFLD is associated with obesity and insulin resistance. There is increase in lipolysis in adipose tissue and increased flux of free fatty acids (FFAs) to the liver. There is also increased hepatic de novo lipogenesis in the presence of insulin resistance and hyperinsulinemia. In response to increased FFA delivery to the liver, fat oxidation and ketogenesis are accelerated. However, these mechanisms are unable to compensate for chronic nutrient overload to the liver cells. Fatty acid oxidation eventually becomes insufficient and incomplete leading to accumulation of triglycerides (steatosis) and toxic lipid intermediates such as ceramides and diacylglycerol. These lipid intermediates further increase insulin resistance. Hepatic gluconeogenesis is stimulated leading to overactivity of tricarboxylic acid cycle to provide intermediates for gluconeogenesis. The generation of NADH and $FADH_2$ is significantly increased causing increased electron flux through the electron transport chain. This leads to overproduction of reactive oxygen species (ROS) and oxidative stress. Increased ROS production and consequent oxidative stress are considered important mediator of hepatic inflammation, fibrosis, and eventual progression to cirrhosis and hepatocellular carcinoma.

CONCLUSIONS

The role of mitochondria in the pathogenesis of NAFLD is not well defined. However, the inability of mitochondrial fat oxidation to keep up with nutrient overload contributes to hepatic steatosis, and overproduction of ROS leading to oxidative stress plays a significant role in the progression of simple steatosis to NASH and liver cirrhosis. Interventions to decrease fat overload to liver such as weight loss and dietary changes, stimulating hepatocyte mitochondrial biogenesis by either lifestyle changes such as exercise or medication such as Pioglitazone, and decreasing oxidative stress by antioxidants are main proven or experimental approaches to prevent and treat NAFLD.

Mitochondrial Medicine. https://doi.org/10.1016/B978-0-12-817006-9.00018-6

SUGGESTED READING

1. Hashimoto E, Taniai M, Tokushige K. Characteristics and diagnosis of NAFLD/NASH. *J Gastroenterol Hepatol.* 2013; 28:64–70.

2. Sunny NE, Bril F, Cusi K. Mitochondrial adaptation in nonalcoholic fatty liver disease: novel mechanisms and treatment strategies. *Trends Endocrinol Metabol.* 2017;28: 250–260.

CHAPTER 19

Mitochondria and Cardiovascular Diseases

ABSTRACT

Cardiovascular diseases are a major cause of morbidity and mortality worldwide. The role of mitochondria in cardiovascular physiology and the pathogenesis of cardiovascular diseases is increasingly being recognized. Given the magnitude and public health burden of cardiovascular diseases, targeting mitochondrial dysfunction provides an exciting opportunity.

KEYWORDS

Atherosclerosis; Heart failure; Mitochondrial dysfunction; Oxidative stress; Stroke.

CARDIOVASCULAR DISEASES

Cardiovascular diseases (CVDs) are diseases of heart and blood vessels. It generally refers to diseases caused by the narrowing of blood vessels (atherosclerosis). The common CVDs are angina, myocardial infarction, and stroke. Other CVDs are cardiomyopathy and heart failure. CVDs are a major public health burden and the leading cause of death in the world. According to the Center for Disease Control and Prevention (CDC), every 1 in 4 death in the United States is related to CVD.

Improvements in treatment have led to the decline of CVD mortality rates over the past two decades. However, it remains the leading cause of death. A better understanding of the molecular pathogenesis of CVD is needed to find novel and complementary therapies. The role of mitochondria and mitochondrial dysfunction in the pathogenesis of CVD is being increasingly recognized. Targeting mitochondrial dysfunction may provide a complementary approach to the existing therapies to decrease morbidity and mortality from CVD.

ATHEROSCLEROSIS

Atherosclerosis is characterized by the deposit of fatty plaques in the walls of large and medium arteries. It is a chronic inflammatory disease and the main underlying cause of CVD. It is initiated by deposition of low-density lipoprotein (LDL) particles in the subendothelial space of arteries. The accumulated LDL gets modified to oxidized LDL. The hyperlipidemia and proinflammatory condition associated with metabolic syndrome favor these processes. There is upregulation of adhesion molecules on endothelium in response to the accumulated oxidized LDL leading to monocyte adhesion and transmigration into subendothelial space. Monocytes then transform to macrophages and engulf oxidized LDL to become foam cells. Accumulation of oxidized LDL in foam cells leads to oxidative stress and release of proinflammatory cytokines. These lead to apoptosis and necrosis of foam cells and migration of smooth muscle cells (SMC) from the arterial wall to subendothelial space. The migrated SMC localize between endothelium and necrotic foam cells. Smooth muscle cells secrete extracellular matrix and form a fibrous cap resulting into fibrous plaque. However, over time SMC also accumulates oxidized LDL and transform into foam cells. They undergo apoptosis and necrosis due to oxidative stress leading to thinning of the fibrous cap that becomes prone to rupture by the proteases produced by inflammatory cells. Rupture of fibrous plaque initiates thrombosis and may result in complications such as myocardial infarction and stroke.

Oxidative stress and inflammation are central to the pathogenesis of atherosclerosis. Many of the risk factors associated with atherosclerosis such as metabolic syndrome, insulin resistance, and smoking can cause mitochondrial dysfunction. Dyslipidemia and inflammation associated with metabolic syndrome and insulin resistance cause endothelial dysfunction. Endothelial NADPH oxidase is activated while endothelial nitric oxide synthase is uncoupled leading to increased production of reactive oxygen species (ROS) and reactive nitrogen species (RNS). These together cause mitochondrial dysfunction and a further increase in ROS and oxidative stress. The result is worsening of endothelial dysfunction leading to leukocyte adhesion and migration to subendothelial space, secretion of proinflammatory cytokines, and eventually endothelial cell death. The oxidative stress also causes increased conversion

Mitochondrial Medicine. https://doi.org/10.1016/B978-0-12-817006-9.00019-8

of LDL to oxidized LDL that is then engulfed by macrophages leading to the perpetuation of inflammation. Smoking can damage mitochondrial DNA and propagate oxidative stress by activating NADPH oxidase. Mitochondrial dysfunction whether caused by increased ROS, RNS, or DNA damage accelerates the oxidative stress by increased ROS production leading to a vicious cycle. In addition, mitochondrial dysfunction leads to inflammation by activation of the NLRP3 inflammasome which converts the proinflammatory cytokine, interleukin 1β precursor to its mature form. Mitochondrial dysfunction also leads to a decrease in ATP formation, loss of mitochondrial membrane potential, the opening of mitochondrial permeability transition pore (MPTP) and release of mitochondrial DNA into the cytoplasm. Increased ROS and mitochondrial DNA can activate NLRP3 resulting in inflammation. The prolonged opening of MPTP leads to apoptosis. Thus, mitochondrial dysfunction is related to initiation as well as propagation of atherosclerotic lesions by promoting oxidative stress, inflammation, and cell death and may be useful therapeutic target.

MYOCARDIAL INFARCTION

The heart is metabolically most active organ. About 90% of energy consumed by heart is provided by the mitochondrial respiratory chain. Mitochondria occupy about 30% of the cell volume of cardiac myocytes. Hence, the function of the heart is heavily dependent on mitochondria and mitochondrial dysfunction plays key roles in pathological heart conditions. During myocardial infarction, ischemia leads to progressive decline in ATP production by mitochondria due to lack of oxygen and substrates. This leads to intracellular acidosis and a rise in intracellular Na^+ due to activation of Na^+/H^+ exchanger at the cell membrane. Intracellular Na^+ also increases due to inhibition of ATP-dependent Na^+/K^+ exchanger on the cell membrane. Increased intracellular Na^+ leads to increased intracellular calcium due to activation of Na^+/Ca^{2+} exchanger on the cell membrane. Intracellular calcium also increases due to diminished activity of ATP-dependent sarcoplasmic reticulum (SR) calcium channels which sequester calcium in SR. A rise in cytoplasmic calcium leads to increased intramitochondrial calcium. The rise in intramitochondrial calcium is facilitated by reversal of mitochondrial sodium calcium exchanger (mNCX) function that normally causes efflux of calcium from the mitochondria in exchange of calcium. In persistent ischemia, ATP depletion and rise in intramitochondrial calcium lead to the opening of MPTP, mitochondrial swelling, and shut down resulting in cardiomyocyte necrosis. In surviving cardiomyocytes, reperfusion leads to enhanced production of ROS. The increased ROS in the presence of increased intramitochondrial calcium and lack of ATP lead to the opening of MPTP and myocyte death. Thus, mitochondria play a key role in myocyte injury and death during ischemia-reperfusion. The novel protective strategies to salvage cardiomyocyte from ischemia-reperfusion injury include compounds to block MPTP and inhibit mNCX during ischemia-reperfusion to prevent intramitochondrial calcium overload.

CARDIOMYOPATHY AND ARRHYTHMIA

Cardiomyopathy is defined as a disease of myocardium where there is a structural or functional abnormality of the heart muscle. It can lead to heart failure, arrhythmia, and sudden death. There are three main types of cardiomyopathy—dilated cardiomyopathy (DCM), hypertrophic cardiomyopathy (HCM), and restrictive cardiomyopathy (RCM). Dilated cardiomyopathy is the most common type and is the main cause of heart failure. Cardiomyopathy can be primary (for example, caused by a mutation in genes for muscle chain), secondary (infection, autoimmunity, ischemia, toxins, pressure, or volume overload), or idiopathic. The heart is one of the main organs affected in mitochondrial diseases. Cardiomyopathy is sometimes the main manifestations of mitochondrial disease. Table 19.1 outlines some of the mitochondrial diseases commonly associated with cardiomyopathy and arrhythmia. However, cardiac evaluation should be done routinely in any mitochondrial disease.

HEART FAILURE

Heart failure can be defined as the inability of the heart muscles to pump enough blood to meet the metabolic needs of the body. It can be caused by ischemic damage of the heart muscle from the myocardial infarction or other conditions such as hypertension, cardiomyopathy, or valvular heart disease. A decrease in heart pump function leads to compensatory adrenergic stimulation to maintain cardiac output. Although the adrenergic stimulation is beneficial in short term, persistent adrenergic stimulation can further deteriorate cardiac function. The increased workload on cardiomyocyte leads to stimulation of mitochondrial oxidative phosphorylation and increased flux of electrons through the electron transport chain (ETC) leading to an increase in the leak of electron and excessive ROS generation. Excessive ROS further damages ETC complexes and sarcomere proteins leading

TABLE 19.1
Mitochondrial Diseases Commonly Associated With Cardiomyopathy or Arrhythmia

Mitochondrial Disease	Gene	Inheritance Pattern	Main Cardiac Manifestation
Barth syndrome	*TAZ*	X-linked	DCM, HCM (rare), noncompaction cardiomyopathy
Coenzymeq10 deficiency	*COQ2, COQ4, COQ9*	Autosomal recessive	HCM
Mitochondrial complex V deficiency	*TMEM70*	Autosomal recessive	HCM
Friedreich's ataxia	*FXN*	Autosomal recessive	HCM
MERRF (myoclonic epilepsy with ragged-red fibers) syndrome	*MT-TK*	Maternal	DCM
MELAS (mitochondrial encephalomyopathy, lactic acidosis, and stroke-like episodes) syndrome	*MT-TL1*	Maternal	HCM
Kearns–Sayre Syndrome	Mitochondrial DNA deletion	Sporadic	Heart block, arrhythmia

to a decrease in contractility. Current therapy of heart failure involves reducing the workload on the heart (by vasodilation) and decreasing heart rate thus reducing myocardial oxygen consumption and attenuating the harmful effect of adrenergic stimulation. β-blockers are one of the main lines of therapy for heart failure. However, there is a limitation of this approach as beyond a point intervention on these lines will lead to hypotension and bradycardia. Hence, novel therapies to address the underlying pathophysiology of heart failure are needed. One approach could be mitochondria-targeted antioxidants to prevent sarcomere injury and ETC impairment. Cardiolipin is an integral component of inner mitochondrial membrane and important for ETC function as it holds the individual ETC complexes together as a supercomplex. Cardiolipin is preferentially damaged by ROS. Bendavia (MTP-131), which stabilizes cardiolipin, has been shown to be beneficial in reducing cardiac injury in animal models.

STROKE

Cerebrovascular stroke is a major cause of disability and mortality in the adult population. Mitochondria play a significant role in the pathogenesis of brain damage in stroke. During brain ischemia, there is overstimulation of glutamate receptors resulting in calcium overload of neurons. This leads to mitochondrial calcium overload and the opening of MPTP. The opening of MPTP causes loss of mitochondrial membrane potential, osmotic swelling, and shut down of oxidative phosphorylation leading neuronal death. During reperfusion, excessive ROS generation aggravates this process. Surrounding the necrotic core of brain tissue, there is peri-infarct zone or "ischemic penumbra." Neurons in this zone are damaged but can be salvaged by reperfusion. However, they are also at risk of reperfusion injury. Optimization of stroke treatment may maximize the recovery of neurons in this zone. A potential therapeutic approach is inhibition of MPTP since it plays a central role in neuronal death. Targeted antioxidant therapy is another potential therapy to limit ischemia-reperfusion induced brain damage. Inflammation plays a significant role in neuronal injury in ischemia-reperfusion. Mitochondria contribute to initiation and propagation of inflammation. The inflammasome NLRP3 is activated by excessive ROS generated from mitochondrial dysfunction. In addition, mitochondrial DNA when released in the cytoplasm after MPTP opening leads to activation on NLRP3 that converts proinflammatory interleukin 1β precursor to its mature form. Released mitochondrial DNA from neuronal death acts as danger-associated molecular pattern due to its resemblance to the bacterial genome and activates microglia further enhancing inflammation. Inhibition of NLRP3 activation has therapeutic potential in stroke.

SUGGESTED READING

1. Perrotta I, Aquila S. The role of oxidative stress and autophagy in atherosclerosis. *Oxid Med Cell Longev.* 2015;2015: 130315.

2. Yu EP, Bennett MR. Mitochondrial DNA damage and atherosclerosis. *Trends Endocrinol Metabol.* 2014;25: 481−487.

3. Ong SB, Samangouei P, Kalkhoran SB, Hausenloy DJ. The mitochondrial permeability transition pore and its role in myocardial ischemia reperfusion injury. *J Mol Cell Cardiol.* 2015;78:23−34.

4. Brown DA, Perry JB, Allen ME, et al. Expert consensus document: mitochondrial function as a therapeutic target in heart failure. *Nat Rev Cardiol.* 2017;14:238−250.

5. Yang JL, Mukda S, Chen SD. Diverse roles of mitochondria in ischemic stroke. *Redox Biol.* 2018;16:263−275.

CHAPTER 20

Mitochondria and Late-Onset Neurodegenerative Diseases

ABSTRACT

Neurons are very active metabolically and have high energy demand. In addition, the brain is more susceptible to damage from oxidative stress. Mitochondria provide most of the brain's energy need by oxidative phosphorylation and are also the main source of reactive oxygen species. Thus, mitochondrial dysfunction plays a significant role in the pathogenesis of adult-onset neurodegenerative disorders.

KEYWORDS

Alzheimer disease; Amyotrophic lateral sclerosis; Friedreich ataxia; Huntington disease; Late onset; Neurodegenerative disease; Parkinson disease.

ALZHEIMER DISEASE

Alzheimer disease (AD) is the leading cause of dementia worldwide. It is a progressive neurodegenerative disorder characterized histopatholgically by extracellular amyloid plaques and intracellular neurofibrillary tangles. The pathogenesis of AD is not well understood. According to "amyloid cascade hypothesis," amyloid precursor protein (APP) is abnormally processed in AD leading to the formation of amyloid β-peptides that are present as soluble oligomers and insoluble extracellular deposits or amyloid plaques. Soluble amyloid β-peptide oligomers are particularly neurotoxic and lead to synaptic degeneration, neurofibrillary tangles, and neuronal loss.

Mitochondrial dysfunction is consistently observed in AD. The activity of respiratory chain complex IV is reduced in AD. The levels of ROS are increased in AD brain. In addition, there is abnormal mitochondrial axonal transport and dynamics. Many of these changes are seen very early in the disease. Moreover, mitochondrial dysfunction in AD was shown to be directly related to abnormal APP processing. This led to proposal of "mitochondrial cascade hypothesis." According to this hypothesis, mitochondrial dysfunction is the primary cause of AD. Decline in mitochondrial function below a threshold leads to abnormal APP processing, oxidative stress, neurofibrillary tangle formation, and

neuronal death. There is genetic variation in mitochondrial function among individuals. The age-associated decline in mitochondrial function in genetically predisposed individual leads to the decline of mitochondrial function below the threshold to trigger this cascade. However, there are limitations to this hypothesis. One argument against this hypothesis is that patients with known mitochondrial diseases with severe decline in mitochondrial function do not show characteristic clinical or pathological features of AD. Thus, mitochondrial dysfunction seen in AD may be secondary or a bystander. Recently, mitochondria-associated endoplasmic reticulum membrane or "MAM" has been proposed to be a mediator of mitochondrial dysfunction in AD. APP is cleaved by nonamyloidogenic and amyloidogenic pathways. In the latter, APP is first cleaved to a soluble N-terminal and a shorter C-terminal (C99) fragment. C99 is then cleaved at MAM to amyloid β-peptide by γ-secretase. Presenilin-1 (PS1) and presenilin-2 (PS2) are components of γ-secretase. In conditions where the amyloidogenic pathway is favored such as *APP* mutation or C99 processing is deficient (presenilin mutation), there is excessive C99 accumulation at MAM. This causes upregulated sphingomyelinase activity and increased sphingomyelin turnover at the MAM leading to increased ceramide production. Increased ceramide destabilizes mitochondrial ETC supercomplex causing mitochondrial dysfunction, oxidative stress, neurofibrillary tangle formation, and ultimately neuronal death. According to this hypothesis, mitochondrial dysfunction is a mediator of AD but not a primary event. Mitochondrial dysfunction induced by accumulated C99 may explain the early onset AD in familial cases associated with *APP*, *PS1*, and *PS2* mutations. Overall, mitochondrial dysfunction is an important component of AD pathogenesis and a potential therapeutic target.

PARKINSON DISEASE

Parkinson disease (PD) is the second most common neurodegenerative disorder of adulthood. The characteristic pathological features of PD are an intracellular

Mitochondrial Medicine. https://doi.org/10.1016/B978-0-12-817006-9.00020-4

accumulation of α-synuclein in neurons and loss of neurons in substantia nigra pars compacta (SNpc). Mitochondrial dysfunction, particularly complex I deficiency is associated with PD. Exposure to the compound 1-methyl-4-phenyl-1, 2, 3, 4-tetrahydropyridine (MPTP) leads to Parkinsonism. MPTP is a complex I inhibitor and selectively kills neurons of SNpc. Complex I deficiency leads to neuronal damage by a decrease in ATP production and increased ROS generation. Dopaminergic neurons of SNpc are particularly vulnerable to damage. They have long poorly myelinated axons. Mitochondria are fissioned into small punctate mitochondria to able to be transported to the synaptic terminals. These mitochondria are more susceptible to complex I inhibition and damage from ROS. In addition, these neurons are autonomously active driven by voltage-dependent L-type calcium channels. There is a sustained rise in intracellular calcium that leads to mitochondrial calcium overload. Mitochondrial calcium overload causes decreased mitochondrial respiration, increased ROS generation, and lower threshold for opening of MPTP predisposing to neuronal death. Thus, mitochondrial dysfunction has an important role in the pathogenesis of PD. Autosomal recessive forms of PD are caused by mutations in *PARK2* and *PINK1*. PARK2 and PINK1 work in coordinated fashion to remove damaged mitochondria by mitophagy. Mitophagy is important to maintain healthy mitochondrial pool in cells. A defect in mitophagy leads to accumulation of damaged mitochondria, mitochondrial dysfunction, increased ROS generation, and neuronal loss.

PD is characterized by intraneuronal α-synuclein accumulation. It is normally present in cells as soluble monomer form. However, due to excessive production (α-synuclein gene mutation or polymorphism) or inadequate clearance (aging, defects of lysosomal autophagy or ubiquitin-proteasome degradation pathways), it accumulates in PD. The α-synuclein monomers aggregate to form toxic soluble oligomers and ultimately insoluble fibrils. Oligomers are toxic to mitochondria. They can bind with cardiolipin causing mitochondrial dysfunction and increased ROS generation. Increased ROS can aggravate misfolding of α-synuclein and aggregation thus resulting in a vicious cycle.

AMYOTROPHIC LATERAL SCLEROSIS

Amyotrophic lateral sclerosis (ALS) is a progressive neurodegenerative disorder characterized by loss of upper and lower motor neurons resulting in gradual paralysis, muscle atrophy, and respiratory failure. Motor neurons cannot maintain its axonal projection leading to axonal retraction and denervation. The reason for selective damage of motor neurons is unknown but their large size and need to maintain long axonal projections may make them vulnerable to metabolic abnormalities. Because neurons are highly dependent on mitochondria for energy supply, mitochondria are essential for neuronal function and survival and mitochondrial dysfunction is a common theme in neurodegenerative diseases including ALS. About 10% of ALS cases are caused by familial, mostly autosomal-dominant mutations. Mutation in *SOD1* gene related to ALS leads to mitochondrial dysfunction as the mutant SOD1 protein aggregates in the mitochondrial intermembranous space and impairs mitochondrial protein import and ETC function. Abnormal mitochondrial morphology, decreased respiratory chain function, impairment of mitochondrial membrane potential, oxidative stress, impaired mitochondrial axonal transport, and decreased mitochondrial calcium buffering capacity are frequently associated with familial and sporadic forms of ALS. In familial form, mitochondrial dysfunction is secondary to the causative mutation. However, in sporadic forms, it may be secondary to other dysregulated pathways seen in ALS, such as abnormal proteostasis, aberrant RNA processing, and glutamate excitotoxicity.

FRIEDREICH ATAXIA

Friedreich ataxia (FA) is adolescent-onset progressive neurodegenerative condition and the most common form of inherited ataxia. It is characterized by progressive ataxia, cardiomyopathy, and diabetes. There is progressive neuronal loss and atrophy, primarily in dorsal root ganglia of the spine and dentate nucleus of the cerebellum. FA is an autosomal recessive disorder caused by homozygous GAA repeat extension mutation in the *FXN* gene. Frataxin level is reduced in FA. Frataxin is a mitochondrial matrix protein and is essential for iron—sulfur-complex (ISC) protein assembly. Mitochondrial ETC complexes I, II, and III contain ISC. Therefore, frataxin plays an essential role in ETC assembly and function. Frataxin deficiency also leads to reduced intramitochondrial heme and ferritin synthesis. There is less efflux of heme and ISC proteins from the mitochondria and simultaneously mitochondrial iron uptake is enhanced resulting in mitochondrial iron overload. Abnormal ETC function and iron overload together lead to excessive ROS generation and consequent damages to lipids, DNA, and proteins.

HUNTINGTON DISEASE

Huntington disease (HD) is characterized by choreiform movements, dementia, and psychiatric disturbances. It is caused by expansion of CAG trinucleotide repeat on huntingtin gene. HD is inherited in an autosomal dominant manner. Mutant huntingtin protein has abnormally long polyglutamine sequence that leads to their oligomerization and aggregation. The mutant huntingtin aggregates are toxic to neurons and cause neuronal dysfunction and death by several mechanisms including mitochondrial dysfunction. Deficiency of ETC complex II is seen in HD, and complex II inhibition in a mouse model has been shown to reproduce a histological and neurochemical picture of HD. Mutant huntingtin localizes to mitochondria and can accelerate the mitophagy process by binding to valosin-containing protein (VCP) which then binds with autophagosome. Mutant huntingtin can also cause excessive mitochondrial fragmentation by activating dynamin-related protein-1 (DRP-1) on mitochondria. Thus, there is reduced mitochondrial content and accumulation of fragmented mitochondria. Mutant huntingtin also increases mitochondrial calcium influx leading to mitochondrial calcium overload and the opening of MPTP that may lead to neuronal apoptosis. The disturbance in calcium homeostasis also impairs mitochondrial transport in the axons. Moreover, mutant huntingtin localizes to the nucleus and inhibits the expression of PGC1α causing reduced mitochondrial biogenesis.

CONCLUSIONS

Neurons are metabolically highly active and depend on mitochondria for ATP supply for vital functions such as neurotransmission. In addition, mature neurons are terminally differentiated and cannot be replaced. Mitochondrial dysfunction can cause neuronal dysfunction and permanent neuronal loss. Thus, mitochondrial dysfunction is central to the pathogenesis of late-onset neurodegenerative diseases.

SUGGESTED READING

1. Swerdlow RH, Burns JM, Khan SM. The Alzheimer's disease mitochondrial cascade hypothesis: progress and perspectives. *Biochim Biophys Acta*. 2014;1842:1219−1231.
2. Area-Gomez E, de Groof A, Bonilla E, Montesinos J, Tanji K, Boldogh I, Pon L, Schon EA. A key role for MAM in mediating mitochondrial dysfunction in Alzheimer disease. *Cell Death Dis*. 2018;9:335.
3. Ferreira M, Massano J. An updated review of Parkinson's disease genetics and clinicopathological correlations. *Acta Neurol Scand*. 2017;135:273−284.
4. Shi P, Gal J, Kwinter DM, Liu X, Zhu H. Mitochondrial dysfunction in amyotrophic lateral sclerosis. *Biochim Biophys Acta*. 2010;1802:45−51.
5. Koeppen AH. Friedreich's ataxia: pathology, pathogenesis, and molecular genetics. *J Neurol Sci*. 2011;303:1−12.
6. Bossy-Wetzel E, Petrilli A, Knott AB. Mutant huntingtin and mitochondrial dysfunction. *Trends Neurosci*. 2008;31:609−616.

CHAPTER 21

Mitochondria and Cancer

ABSTRACT

Cancer can be defined as an uncontrolled proliferation of cells that can invade local tissue and spread to distant tissues. Resistance to cell death, metabolic rewiring to sustain uncontrolled proliferation, stromal remodeling, and phenotypic plasticity to facilitate metastasis are a few of the characteristics of carcinogenesis. By virtue of their bioenergetic, biosynthetic, and redox hemostasis roles, mitochondria are increasingly being recognized as important mediators of carcinogenesis and potential therapeutic target.

KEYWORDS

Cancer; Innate immunity; Metastasis; Mitochondria; Oncometabolite.

CARCINOGENESIS

Carcinogenesis can be divided into three stages—initiation, promotion, and progression. Initiation (neoplastic transformation) is the process where a normal cell converts to a neoplastic precursor by acquiring replicative immortality. Promotion is an uncontrolled proliferation of the transformed cell. Tumor progression is associated with local invasion and distant metastasis. In the following section, the role of mitochondria in each of these stages of carcinogenesis is outlined.

ROLE OF MITOCHONDRIA IN NEOPLASTIC TRANSFORMATION

Mitochondria play a key role in apoptosis. In the presence of apoptotic stress such as DNA damage, tumor suppressor proteins such as p53 activate the apoptotic pathway in damaged cells to prevent cells from becoming a neoplastic precursor. In normal cells, p53 activates proapoptotic proteins BAX and BAK at the mitochondrial membrane leading to mitochondrial outer membrane permeabilization (MOMP), release of cytochrome c, and activation of caspase in the cytoplasm resulting in cell death. Cancer cells evade apoptosis by either loss of tumor suppressors such as p53 or upregulation of antiapoptotic bcl-2 proteins that prevent MOMP. Thus, cells with DNA damage evade apoptosis and accumulate genetic damage to transform into a neoplastic precursor.

Tumor initiation can also be due to the activation of proto-oncogenes. Excessive reactive oxygen species (ROS) generated in mitochondria can not only cause conversion of proto-oncogenes to oncogenes but can also trigger the activation of oncogenic signal pathways such as epidermal growth factor signaling pathway. Mitochondrial dysfunction can also lead to oncogenesis by the accumulation of certain tricarboxylic acid cycle intermediates such as succinate, fumarate, and 2-hydroxyglutarate that act as oncometabolites. Accumulation of these compounds leads to change in the gene expression profile of cell and activation of the oncogenic transcriptional program.

ROLE OF MITOCHONDRIA IN TUMOR PROLIFERATION

After neoplastic transformation, cancer cells undergo uncontrolled proliferation and clonal expansion. This is associated with very high metabolic demand and often hypoxia and nutrition-poor environment. Cancer cells rewire their metabolism to meet these challenges. Most cancer cells switch to predominant glycolysis for energy supply (also called Warburg effect). This not only ensures ATP supply but also provides intermediates for lipid and amino acid synthesis required for uncontrolled proliferation. Activated glycolysis provides building blocks for nucleotide synthesis and NADPH for fatty acid synthesis and redox hemostasis via the pentose phosphate pathway. Although glycolysis provides most of the ATP for most cancer types, mitochondria are still required for the proliferation of cells. TCA cycle anaplerosis is used for lipid and protein synthesis required for the proliferation. TCA cycle intermediate citrate is cleaved by citrate lyase in the cytoplasm to generate acetyl coenzyme A and oxaloacetate. Acetyl coenzyme A is

Mitochondrial Medicine. https://doi.org/10.1016/B978-0-12-817006-9.00021-6

then used for lipid synthesis. The loss of TCA cycle intermediate is compensated by glutamine that is converted to α-ketoglutarate by glutaminolysis. Thus, mitochondria play an essential role in metabolic reprogramming of the cancer cells to sustain proliferation.

ROLE OF MITOCHONDRIA IN TUMOR PROGRESSION

Tumor progression is associated with more aggressive tumor behavior leading to local invasiveness and distant metastasis. Mitochondrial ROS production is instrumental in the facilitation of these processes. Hypoxia-induced factor 1α (HIF-1α) is stabilized by ROS. HIF-1α leads to upregulation of matrix metalloproteinases that cause connective tissue degradation and remodeling leading to tumor invasiveness. HIF-1α also upregulates vascular endothelial growth factor leading to neoangiogenesis and facilitation of metastasis. Additionally, mitochondria are required for tumor cell migration and metastasis. Mitochondria are concentrated in the leading edge of the migrating cells. Mitochondrial fission is important for this process.

MITOCHONDRIA AND ANTITUMOR INFLAMMATION

Mitochondrial DNA, ATP, and N-formyl peptide released from dying cancer cells behave as damage-associated molecular patterns and activate inflammation. Additionally, mtDNA is released from mitochondria intracellularly under various stresses and activate NLRP3 inflammasome and stimulator of interferon genes (STING), resulting in secretion of interleukin and interferon, respectively. Thus, mitochondria may have an important role in antitumor immune response.

MITOCHONDRIA AND ANTITUMOR THERAPY

Mitochondria are instrumental in apoptosis induced cell death. Many cancer cells evade apoptosis by upregulation of antiapoptotic protein such as bcl2 that binds and sequesters proapoptotic proteins so that it cannot reach mitochondrial membrane and activate BAK and BAX mediate apoptosis. Anticancer agent Venetoclax displaces proapoptotic proteins from bcl2.

These displaced proapoptotic proteins can then associate with BAX and BAK and induce apoptosis of cancer cells. Moreover, cancer cells acquire resistance to therapy through metabolic reprogramming. For example, cancer primarily relying on glycolysis switches to oxidative phosphorylation (OXPHOS) to acquire resistance to therapies that target glycolysis. Thus metabolic rewiring of cancer cells can lead to resistance. Understanding these compensatory metabolic rewiring can lead to novel synergistic therapies. For example, electron transfer chain inhibitory agents can be used as synergistic therapy for cancers that switch to predominantly OXPHOS to acquire resistance. The role of mitochondria in antitumor immunity may be harnessed to produce a vaccine against cancer.

CONCLUSIONS

Mitochondria play key roles in neoplastic transformation, proliferation, and metastasis. Role of mitochondria in apoptosis, bioenergetics, and biosynthesis of macromolecules underlie the dependence of cancer cells on mitochondria during each step of carcinogenesis. Moreover, mitochondria play significant roles in anticancer immunity and resistance to anticancer agents. Understanding the diverse roles of mitochondria in carcinogenesis is very important to devise novel anticancer therapies.

SUGGESTED READING

1. Li C, Zhang G, Zhao L, Ma Z, Chen H. Metabolic reprogramming in cancer cells: glycolysis, glutaminolysis, and Bcl-2 proteins as novel therapeutic targets for cancer. *World J Surg Oncol.* 2016;14:15.
2. Zhao J, Zhang J, Yu M, et al. Mitochondrial dynamics regulates migration and invasion of breast cancer cells. *Oncogene.* 2013;32:4814—4824.
3. Weinberg SE, Sena LA, Chandel NS. Mitochondria in the regulation of innate and adaptive immunity. *Immunity.* 2015;42:406—417.
4. Mihalyova J, Jelinek T, Growkova K, Hrdinka M, Simicek M, Hajek R. Venetoclax: a new wave in hematooncology. *Exp Hematol.* 2018;61:10—25.
5. Morandi A, Indraccolo S. Linking metabolic reprogramming to therapy resistance in cancer. *Biochim Biophys Acta.* 2017;1868:1—6.

Useful Websites

There are several online resources dedicated to mitochondrial diseases and mitochondrial medicine. Following are the few most useful websites.

United Mitochondrial Disease Foundation (www.umdf.org): A very useful website for patients, families, clinicians, and researchers. Contains information about mitochondria and mitochondrial disease in simple language easily understandable by patients or their families. Additionally, ongoing clinical trials are listed.

Wellcome Trust Centre for Mitochondrial Research Newcastle UK (www.newcastle-mitochondria.com): An excellent source of information on mitochondrial diseases for patients, clinicians, and researchers. Contains useful clinical guidelines for management of mitochondrial disease patients.

Mitochondrial Care Network (www.mitonetwork.org): The mitochondrial care network represents a group of physicians at medical centers across the United States who have expertise and experience in providing coordinated, multidisciplinary care for patients with genetic mitochondrial disease.

North American Mitochondrial Disease Consortium (www.rarediseasesnetwork.org/cms/namdc): North American mitochondrial disease consortium (NAMDC) was established to create a network of all clinicians and clinical investigators in North America (US and Canada, with the hope of including Mexico in the future) who follow sizable numbers of patients with mitochondrial diseases and are involved or interested in mitochondrial research. The NAMDC has created a clinical registry for patients, in the hopes of standardizing diagnostic criteria, collecting important standardized information on patients, and facilitating the participation of patients in research on mitochondrial diseases.

MITOMAP (www.mitomap.org): A compendium of polymorphisms and mutations in human mitochondrial DNA.

MSeqDR (https://mseqdr.org): MSeqDR is mitochondrial disease sequence data resource, a centralized and comprehensive genome and phenome bioinformatics resource built by the mitochondrial disease community to facilitate clinical diagnosis and research investigations of individual patient phenotypes, genomes, genes, and variants.

Commonly Prescribed Mitochondrial Supplements

Medication/Supplement	Dose (Pediatric)	Dose (Adult)	Adverse Effects
Coenzyme Q10 (It is available in reduced form called ubiquinol and in oxidized form called ubiquinone. Reduced form has better bioavailability and is preferred)	Ubiquinol 2–8 mg/kg in 2 divided doses Ubiquinone 10–30 mg/kg in 2 divided doses	Ubiquinol 60–600 mg daily in 2 divided doses Ubiquinone 300–2400 mg in 2–3 divided doses	Co Q10 is generally safe. May cause stomach upset, diarrhea, or insomnia. May reduce the efficacy of Warfarin. Safety during pregnancy and breast feeding is not established.
Riboflavin (Vitamin B2)	50–200 mg daily in 2–3 divided doses	50–400 mg daily in 2–3 divided doses	Riboflavin is generally safe. May cause nausea, anorexia, and urine discoloration.
Alpha lipoic acid	25 mg/kg/day	300–600 mg/day	Alpha lipoic acid is generally safe. May decrease blood sugar. Higher dose than recommended should be avoided in children. Safety during pregnancy and breast-feeding is not established.
Vitamin E	1–2 IU/kg/day	100–200 IU daily	Vitamin E in recommended doses is generally safe. May cause stomach upset, diarrhea, headache, or blurred vision. May increase risk of death in people with severe heart condition. May reduce the efficacy of anticoagulants and vitamin K.
Vitamin C	5 mg/kg daily	50–200 mg daily	Vitamin C in recommended doses is generally safe. May cause stomach upset, diarrhea, headache, or insomnia. May increase risk of renal stone. May reduce the efficacy of Warfarin when used in high doses.
L-carnitine	25–100 mg/kg/day in 2–3 divided doses	1000–3000 mg per day in 2–3 divided doses	L-carnitine is generally safe. May cause stomach upset, diarrhea, headache, irritability, or insomnia. May cause fishy body odor when high doses are used.

Continued

Medication/Supplement	Dose (Pediatric)	Dose (Adult)	Adverse Effects
L-creatine	0.1 g/kg daily divided in 3 doses (maximum daily dose 10 g)	2–10 g daily divided in 3 doses	L- creatine is generally safe in recommended doses. May cause stomach upset, diarrhea, muscle cramping, heat intolerance, fever, or dizziness. There is concern for nephrotoxicity when used in high doses.

Index

A

Active phase, Leber hereditary optic neuropathy, 44
Adenine nucleotide translocator (ANT), 60
Adiponectin, 101
Adipose tissue, 101
Aging and age-related disorders
  definition, 97
  DNA mutations and, 97–98
  fusion and fission, 98
  longevity pathways, 99–100
  mitochondrial biogenesis, 98
  mitochondrial oxidative respiration, 97
  mitochondrial proteostasis, 99
  ROS production, 98–99
α-synuclein accumulation, 110
Alpha lipoic acid, 16t–17t
Alzheimer disease (AD), 109
AMP-activated protein kinase (AMPK), 99
Amyloid cascade hypothesis, 109
Amyotrophic lateral sclerosis (ALS), 110
Antiapoptotic protein, 114
Antitumor inflammation, 114
Antitumor therapy, 114
Apoptosis, 3
Arrhythmia, 106
Atherosclerosis, 105–106
Atrophic phase, Leber hereditary optic neuropathy, 44
Autosomal dominant inheritance, 23
Autosomal recessive inheritance, 23

B

Bendavia (MTP-131), 106–107
Biochemical-screening tests, 11t
Brain ischemia, 107–108

C

Calcium homeostasis, 2
Caloric restriction, 99–100
Cancer
  antitumor inflammation, 114
  antitumor therapy, 114
  carcinogenesis, 113
  neoplastic transformation, 113
  tumor progression, 114
  tumor proliferation, 113–114
Carcinogenesis, 113
Cardiolipin, 106–107
  synthesis and remodeling defects, 78t
Cardiomyopathy, 106
Cardiovascular diseases (CVDs)
  atherosclerosis, 105–106
  cardiomyopathy and arrhythmia, 106, 107t
  heart failure, 106
  myocardial infarction, 106
  stroke, 107–108
Cell death, 3
Chorionic villus sampling (CVS), 24
Chronic progressive external ophthalmoplegia (CPEO)
  clinical presentation, 31, 32t
  diagnosis, 31–32
  genetics, 31
  management, 32–33, 32t
Coenzyme Q10, 16t–17t
  deficiency
    clinical presentation, 50, 52t
    management, 53, 54t
    quantification, 13t
Coenzyme Q (CoQ), 1
Cytochrome c, 1
Cytoplasmic salvage pathway, 60

D

Damage-associated molecular pattern (DAMP), 3
Deoxyguanosine kinase (DGK), 59–60
Diabetes mellitus (DM)
  insulin resistance, 101
  oxidative stress, 101–102
  type I and type II, 101
Dichloroacetate (DCA), 90–91
Diet in mitochondrial diseases, 15
Dihydrolipoamide dehydrogenase (DLD)
  activity, 89
  deficiency, 90
Dilated cardiomyopathy (DCM), 106
Dynamin-related protein1 (DRP1), 8, 111
Dyslipidemia, atherosclerosis, 105–106

E

Electron microscopy, 13t
Electron transport chain (ETC), 1, 2f
Enteral feeding, 21
Ethylmalonic encephalopathy (EE), 17t

F

Facial dysmorphism, 89–90
Fe–S cluster biosynthesis, 2–3
Fission 1 homolog protein (FIS1), 8
Frataxin, 110
Friedreich ataxia (FA), 110
Fumarase deficiency, 86t–87t

G

Gene therapy, 19–20
Genetic counseling
  autosomal dominant inheritance, 23
  autosomal recessive inheritance, 23
  mitochondrial DNA mutation, 23–24
  unknown molecular etiology, 24
  X-linked inheritance, 23
Genetic disorders, 5
Genetics
  dynamic interconnected tubular network, 8
  genome, 7
  haplogroups, 9
  heteroplasmy, 7–8
  inheritance, 8–9
  somatic segregation, 7–8
  "threshold effect", 7–8
Genome, 7
Glutamine t-RNA, 68

H

Haplogroups, 9, 45
Heart failure, 106
Hepatic gluconeogenesis, 103
Hereditary leiomyomatosis and renal cell cancer (HLRCC), 87t–88t
Heteroplasmy, 7–8, 20, 45
Histochemistry, 13t
Huntington disease (HD), 111
Hydrogen peroxide, 1–2
Hyperammonemia, 21
Hyperlipidemia, 105
Hypertriglyceridemia, 101
Hypertrophied adipocytes, 101
Hypoinsulinemia, 101
Hypoxia-induced factor 1 α (HIF-1 α), 114

'Note: Page numbers followed by "f" indicate figures and "t" indicate tables.'

I

Impaired mitochondrial fusion, 79t
Inflammation
  atherosclerosis, 105–106
  ischemia-reperfusion, 107–108
  obesity, 101
Inheritance pattern
  DNA maintenance disorder, 65t
  Leigh syndrome (LS), 56t
  mitochondrial DNA maintenance
    disorders, 65t
  mitochondrial homeostasis diseases,
    80t
  pyruvate dehydrogenase (PDH)
    deficiency, 85, 88t
  tricarboxylic acid cycle disorders, 85,
    88t
Inner mitochondrial membrane
  (IMM), 1, 77
Insulin/insulin-like growth factor
  signaling (IIS) pathway, 99
Insulin resistance, 101
Intraoperative management, 21–22
Invasive tissue diagnosis, 12
"Ischemic penumbra", 107–108

K

Kearns-Sayre syndrome (KSS)
  clinical presentation, 31, 32t
  diagnosis, 31–32
  genetics, 31
  management, 32–33, 32t
α-Ketoglutarate dehydrogenase
  (KDH) deficiency, 86t–87t

L

Lactic acidosis, 21
Leber hereditary optic neuropathy
  (LHON), 17t
  clinical presentation, 44
  diagnosis, 45
  genetics, 44–45, 44t
  management, 45, 45t
Leigh syndrome (LS), 31
  clinical presentation, 54–55
  diagnosis, 55–56
  etiological clues, 56t
  genetics, 55, 55t
  inheritance pattern, 56t
  management, 56, 56t
  pyruvate dehydrogenase (PDH)
    deficiency, 89–90
  research trials, 57t
  specific therapies, 57t
Low-density lipoprotein (LDL), 105

M

Magnetic resonance spectroscopy, 14
Mechanistic target of rapamycin
  (mTOR) pathway, 99
Metabolic acidosis, 21
Metabolic syndrome (MetS)
  definition, 101
  oxidative stress, 101–102

Methionine t-RNA formyltransferase
  (MTFMT), 7
1-Methyl-4-phenyl-1, 2, 3, 4-tetrahy-
  dropyridine (MPTP), 109–110
Mitochondria
  functions
    ATP production, 1
    calcium homeostasis, 2
    cell death, 3
    electron transport chain, 1, 2f
    immunity and inflammation, 3
    iron and copper metabolism,
      2–3
    membrane structure, 1
    and reactive oxygen species, 1–2
  genetics. See Genetics
Mitochondrial biogenesis, 18
Mitochondrial cascade hypothesis,
  109
Mitochondrial deletion disorders
  chronic progressive external
    ophthalmoplegia. See Chronic
    progressive external ophthalmo-
    plegia (CPEO)
  Kearns-Sayre syndrome. See Kearns-
    Sayre syndrome (KSS)
  Pearson syndrome. See Pearson
    syndrome
  progressive external ophthalmople-
    gia. See Progressive external
    ophthalmoplegia (PEO)
Mitochondrial diseases
  avoidance of mitotoxic drugs, 15
  clinical features, 5
  clinical spectrum, 5
  diagnosis
    biochemical screening, 11, 11t
    invasive tissue diagnosis, 12
    molecular genetic test, 11–12, 12t
    neuroimaging, 12–14, 13t
  dietary recommendation, 15
  emergency management, 21
  genetic counseling, 23–24
  lifestyle, 15
  mDNA mutations, 27, 27t–28t
  vs. mitochondrial dysfunction, 5
  nuclear gene mutations, 27–29,
    28t–29t
  organ transplantation, 18
  pharmacological therapies, 15,
    16t–17t
  prenatal diagnosis, 24
  reproductive options, 24–25, 24t
  research/experimental therapies
    inhibition of permeability
      transition pore, 19
    mitochondrial biogenesis, 18
    molecular genetic approaches,
      19–20
    nucleosides/nucleotides replenish-
      ing, 19
    restoring abnormal calcium
      homeostasis, 19
    ROS scavenging, 18, 19t

Mitochondrial diseases (Continued)
  surgery, 21–22
  symptomatic therapy, 15–18
Mitochondrial DNA deletion, 23–24
Mitochondrial DNA (mtDNA)
  maintenance disorders
    clinical presentation, 60
    cytoplasmic salvage pathway, 60
    diagnosis, 60, 65t
    genes related, 61t–62t
    inheritance pattern, 65t
    management, 60, 65t
    nuclear genes, 63t–64t
    nucleotide import, 60
    replication approaches, 59
    salvage pathway, 59–60
  point mutation disorders
    clinical presentation, 37
    diagnosis, 37–38
    genes and associated phenotypes,
      39t–40t
    genetics, 37
    management, 38–40, 40t
  replication, 7
  sequencing, 11–12
  transcription, 7
  transcription disorders
    diagnosis, 73–74, 74t
    genetics, 73, 73t
    management, 74–75, 74t
    m-RNA processing, 68, 72t
    nuclear genes, 67, 67t
    pre-RNA processing, 67, 68t
    ribosomal structure and assembly,
      68, 72t
    t-RNA modification, 67–68,
      69t–71t
  translation, 7
Mitochondrial encephalomyopathy,
  lactic acidosis, stroke-like episodes
  (MELAS), 17t
  clinical presentation, 41, 41t
  diagnosis, 37–38, 43
  genetics, 37, 42, 42t
  management, 43, 43t
Mitochondrial fusion, 8
Mitochondrial homeostasis diseases
  diagnosis, 79, 80t
  genetics, 79, 80t
  management, 79, 80t
  mitochondrial quality control
    at cellular level, 78–79
    mitochondrial fusion and fission,
      77–78
    mitochondrial proteome, 77–78
    at molecular level, 78, 79t
    at organellar level, 78, 79t
Mitochondrial membrane integrity,
  77
  diseases, 78t
Mitochondrial membrane potential
  (MMP), 19
Mitochondrial outer membrane
  permeabilization (MOMP), 3, 113

Mitochondrial permeability transition pore (MPTP), 105–106
Mitochondrial sodium calcium exchanger (mNCX) function, 106
Mitochondrial targeting sequence (MTS), 19–20
Mitochondrial termination factor 1 (MTERF1), 7
Mitochondrial translation disorders
  diagnosis, 73–74, 74t
  genetics, 73, 73t
  management, 74–75, 74t
  nuclear genes, 73, 73t
Mitochondrial translation release factor (MTRF1L), 7
Mitohormesis, 99
Mitokines, 99
Mitophagy, 8, 78–79
Mitoribosome, 68, 72t
Molecular genetic test, 11–12, 12t
Multiple cutaneous and uterine leiomyomas (MCUL), 87t–88t
Muscle relaxants, 22
Myocardial infarction, 106
Myoclonic epilepsy, ragged-red fibers (MERRF), 38

N

N-acetyl aspartate (NAA), 14
NADH/CoQ reductase, 20
Neoplastic transformation, 113
Neurodegenerative diseases
  Alzheimer disease (AD), 109
  amyotrophic lateral sclerosis (ALS), 110
  Friedreich ataxia (FA), 110
  Huntington disease (HD), 111
  Parkinson disease (PD), 109–110
Neuroimaging findings, 12–14, 13t
Next-generation sequencing, 11–12
Nod-like receptors (NLR), 3
Nonalcoholic fatty liver disease (NAFLD)
  hepatic de novo lipogenesis, 103
  hepatic fat accumulation, 103
  metabolic syndrome, 103
  obesity and insulin resistance, 103
Nonsyndromic sensorineural hearing loss
  clinical presentation, 46, 46t
  diagnosis, 47
  genetics, 46–47
  management, 47
Nuclear gene mutations, 23, 23t
Nucleotide import, 60

O

Obesity
  definition, 101
  inflammation, 101

Obesity (*Continued*)
  oxidative stress, 101–102
  sedentary lifestyle and excessive nutrition, 101
Oogenesis, 9
Optic atrophy 1 (OPA1), 8
Organ transplantation, 18
Outer mitochondrial membrane (OMM), 1
Oxidative stress, atherosclerosis, 105–106

P

Paraganglioma, 87t–88t
Parkinson disease (PD), 109–110
Pearson syndrome
  clinical presentation, 31, 32t
  diagnosis, 31–32
  genetics, 31
  management, 32–33, 32t
Permeability transition pore (PTP) inhibition, 19
Peroxisome proliferator-activated receptor gamma coactivator 1α (PGC1α), 18
Pheochromocytomas, 87t–88t
Point mutation, 24
Postoperative management, 22
Preimplantation genetic diagnosis (PIGD), 24
Prelingual sensorineural hearing loss, 46
Prenatal diagnosis, 24
Preoperative management, 21
Primary Coenzyme Q10 deficiency, 17t
Programmed cell death, 3
Progressive external ophthalmoplegia (PEO)
  clinical presentation, 31, 32t
  diagnosis, 31–32
  genetics, 31
  management, 32–33, 32t
Propofol, 22
Proteostasis, 99
Pseudouridine synthase (PUS1), 7
Pyruvate carboxylase (PC) deficiency
  benign form, 92
  diagnosis, 93, 94t
  genetics, 92
  infantile form, 92
  management, 93–94, 94t
  neonatal form, 92
  pathogenesis, 92
Pyruvate dehydrogenase (PDH) deficiency
  clinical presentation, 84t–85t, 89–90
  diagnosis, 88t–89t, 90, 91t
  genetics, 90, 91t
  inheritance pattern, 85, 88t
  management, 90–91
  pathophysiology, 84t–85t

Pyruvate metabolism
  aerobic condition, 83
  disorders
    pyruvate carboxylase deficiency. *See* Pyruvate carboxylase (PC) deficiency
    pyruvate dehydrogenase deficiency. *See* Pyruvate dehydrogenase (PDH) deficiency
  oxaloacetate, 83
  schematic representation, 84f

R

Reactive oxygen species (ROS), 1–2
Replication, mtDNA, 7
Respiratory chain, 20
Respiratory chain complex deficiency
  complex I deficiency, 49, 50t
  complex II deficiency, 49, 50t
  complex III deficiency, 49, 51t
  complex IV deficiency, 49, 51t
  complex V deficiency, 49–50, 52t
  CoQ10 deficiency, 50, 52t
  diagnosis, 53, 53t
  Fe–S cluster biogenesis, 50, 52t
  genetic origin, 50t
  genetics, 50, 53t
  management, 53, 54t
Respiratory chain (electron transport chain) enzyme assay, 13t
Riboflavin (Vitamin B2), 16t–17t
Ribonucleotide reductase (RNR), 60

S

Sensorineural hearing loss (SNHL), 46
Sevoflurane, 22
Sirtuin pathway, 99
Skeletal muscle biopsy specimen, 13t
Smooth muscle cells (SMC), 105
Somatic segregation, 7–8
Stroke, 107–108
Succinate dehydrogenase (SDH) deficiency, 86t–87t
Succinyl CoA synthetase (SS)/succinyl CoA ligase deciency, 86t–87t

T

"Threshold effect", 7–8
Thymidine kinase (TK2), 59–60
Thymidine kinase 2 (TK2) deficiency, 19
Transcription, mtDNA, 7
Translation, mtDNA, 7
Tricarboxylic acid (TCA) cycle disorders
  diagnosis, 88t–89t
  inheritance pattern, 85, 88t

Tricarboxylic acid (TCA) cycle
(*Continued*)
    inherited cancer predispositions,
      83, 87t–88t
    inherited disorders, 86t–87t
   schematic representation, 84f
Triheptanoin, 93–94
t-RNA modification, 67–68

Tumor progression, 114
Tumor proliferation, 113–114

V
Valosin-containing protein (VCP),
   111
Venetoclax, 114

Vitamin C, 16t–17t
Vitamin E, 16t–17t
Volatile anesthetics, 22

X
X-linked inheritance, 23

Printed in the United States
By Bookmasters